Copyright © 2018 Marina Vigodsky
All rights reserved.
ISBN: **172759181**
ISBN-13: **978-1727591811**

# 1st Grade Math Comparison, Subtraction and Multiplication Basics

Marina Vigodsky

# Introduction

The 1st Grade Math Comparison, Subtraction and Multiplication Basics workbook introduces 1st graders to three basic arithmetic operations.

Use of a three-sector numbers introduction button for teaching three basic operations provides common contextual background which makes comparison, subtraction and multiplication easier for understanding and practice. From cognitive standpoint a common contextual background turns all three parts of a three-sector number button into one unit which contributes to better understanding of functions of numbers in basic arithmetic operations.

The workbook provides practice for basic arithmetic operations with numbers 1 to 12.

These numbers are equal ( = ).

1 = 1    2 = 2
3 = 3    4 = 4
5 = 5    6 = 6
7 = 7    8 = 8
9 = 9    10 = 10
11 = 11  12 = 12

These numbers are equal ( = ). Fill in the missing numbers.

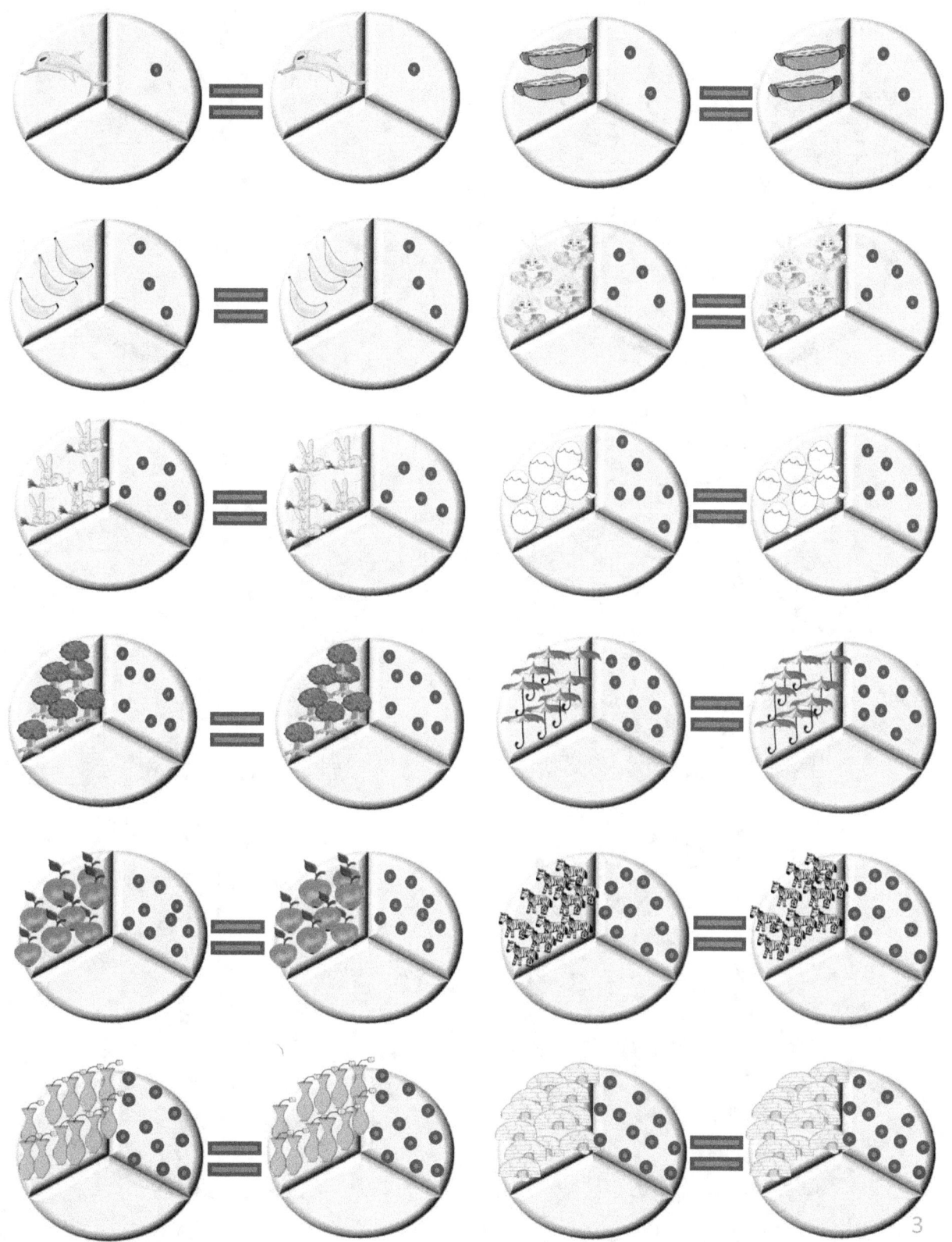

# Comparison. Greater than ( > ).

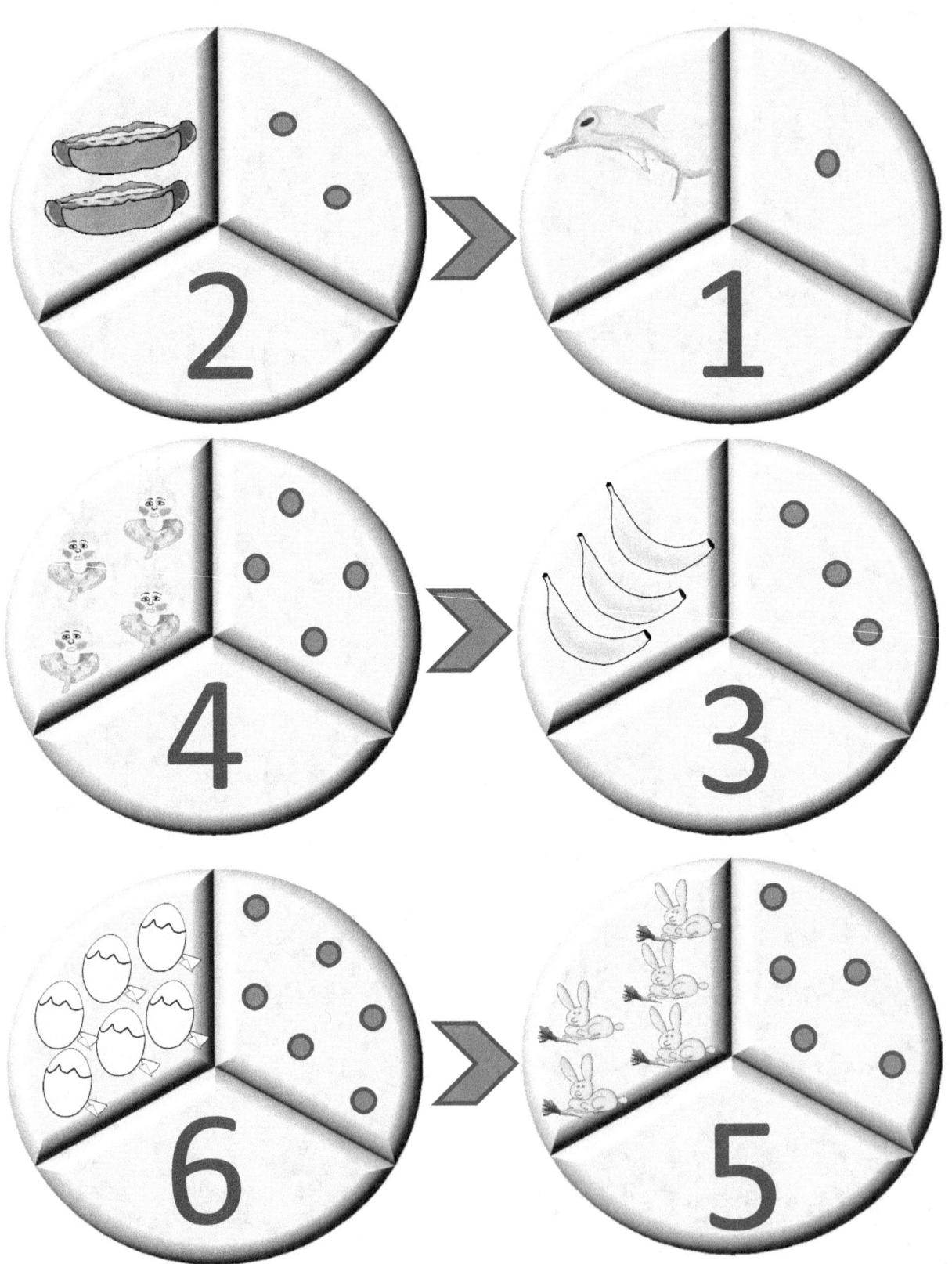

# Comparison. Less than ( < ).

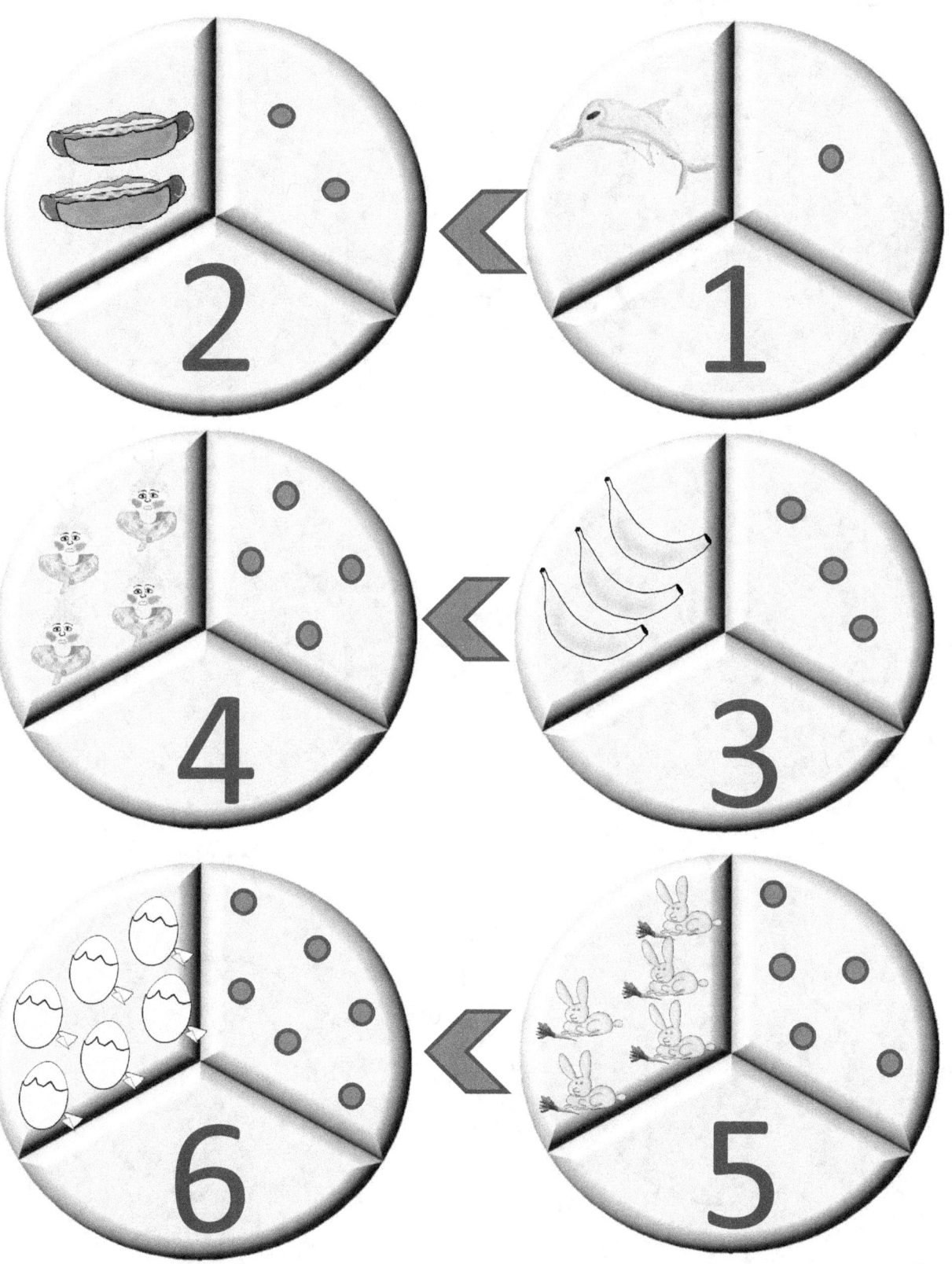

# Comparison. Greater than (>). Fill in the missing numbers.

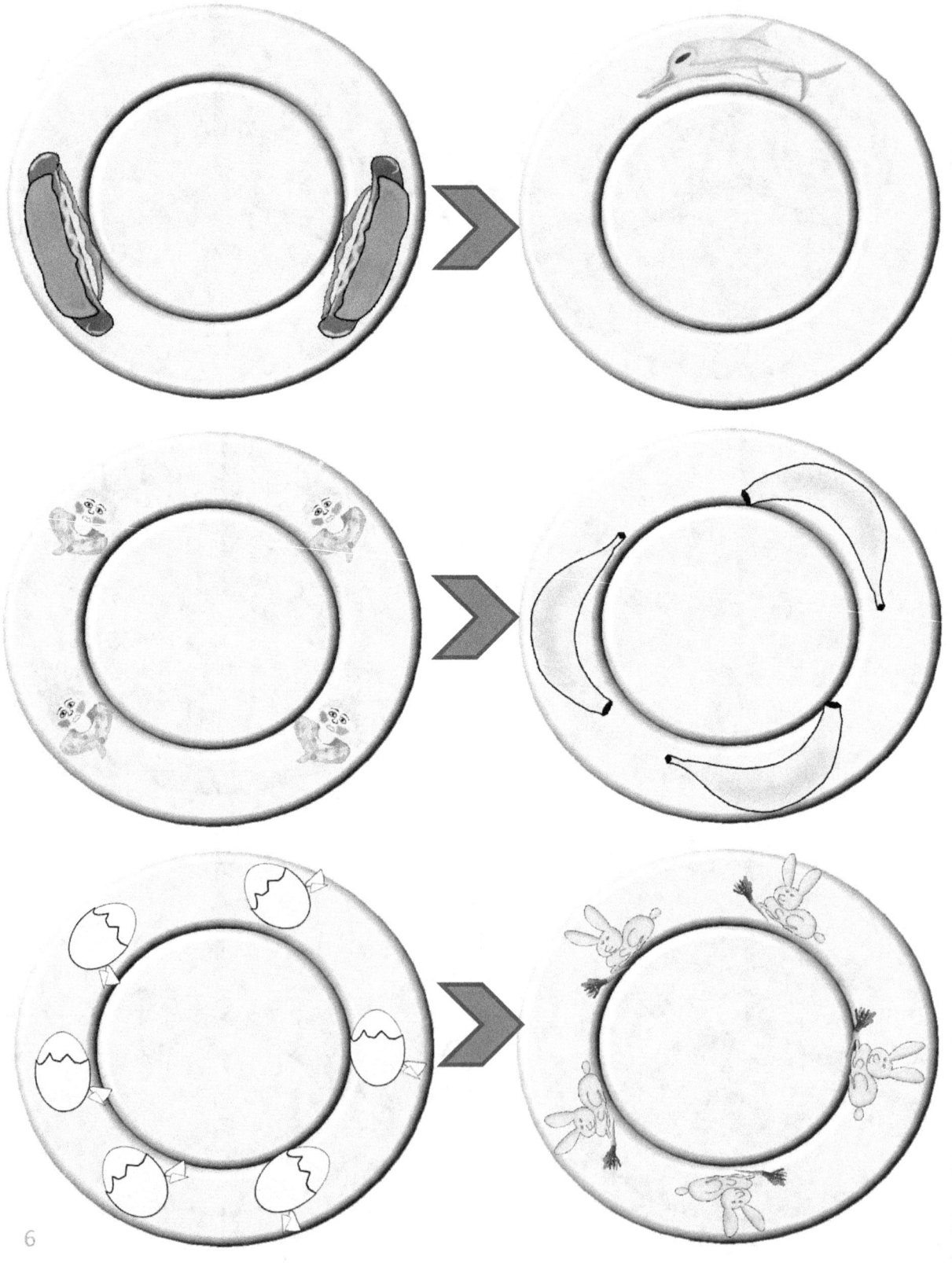

Comparison. Less than (<). Fill in the missing numbers.

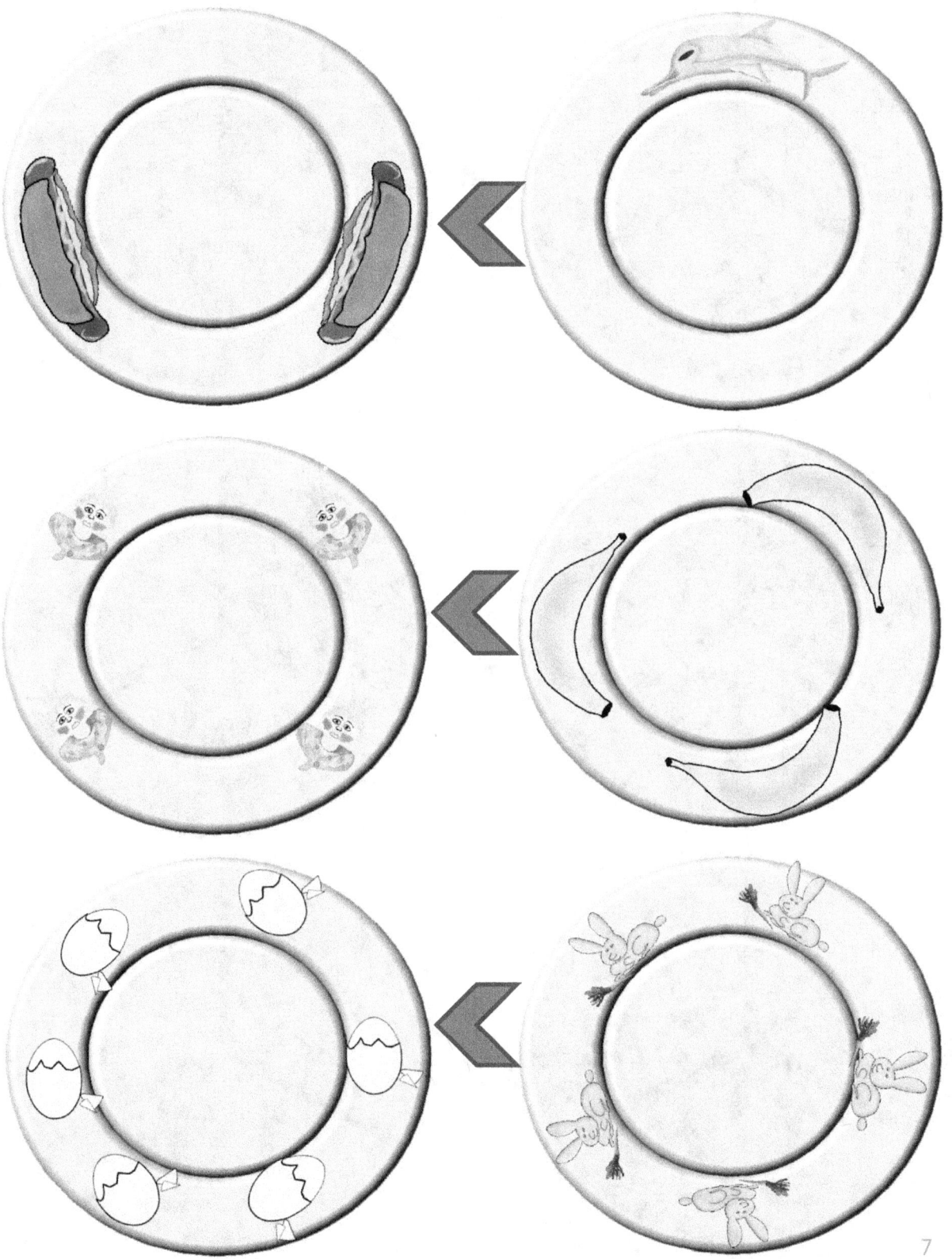

# Comparison. Greater than ( > ).

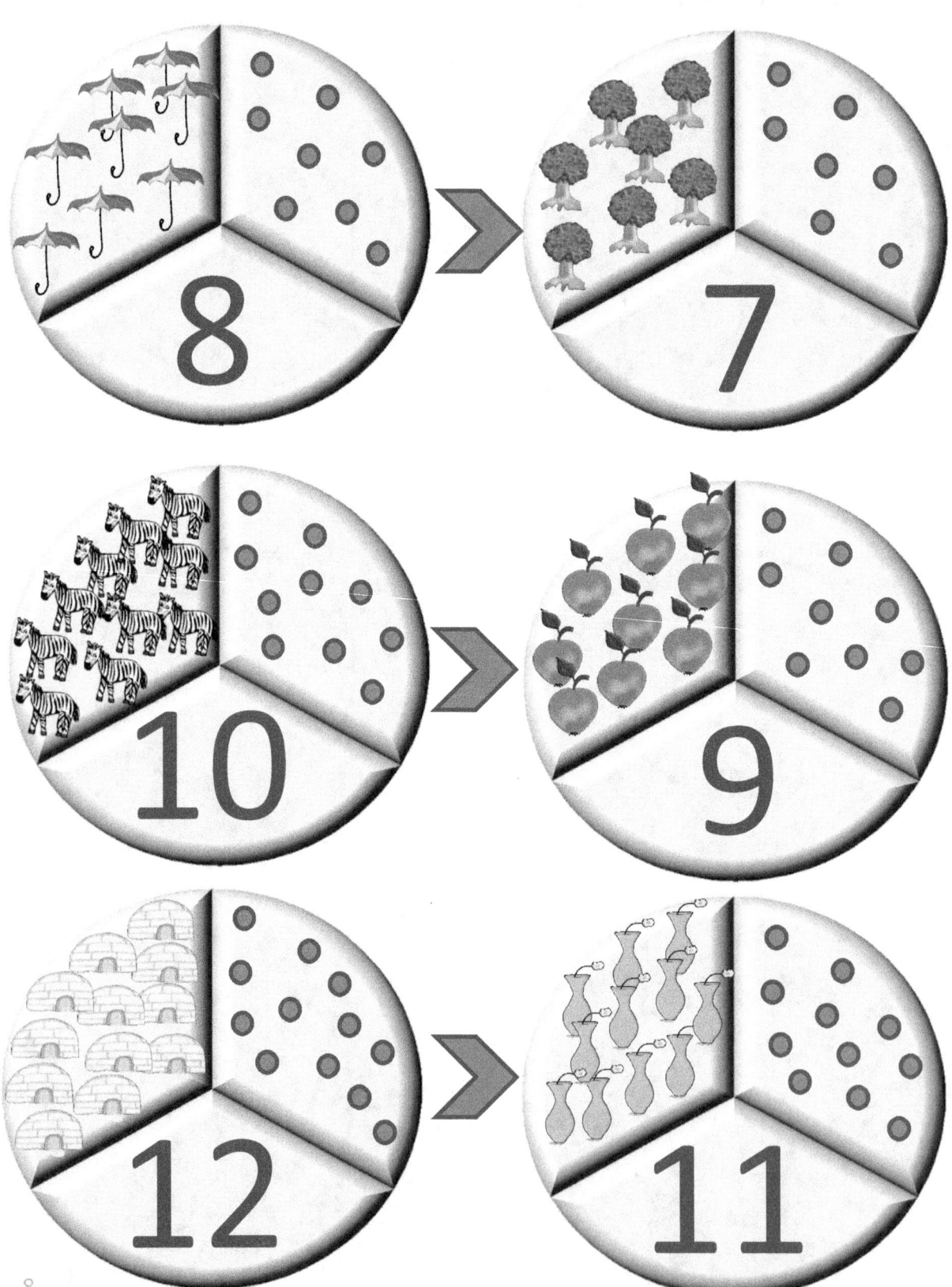

# Comparison. Less than ( < ).

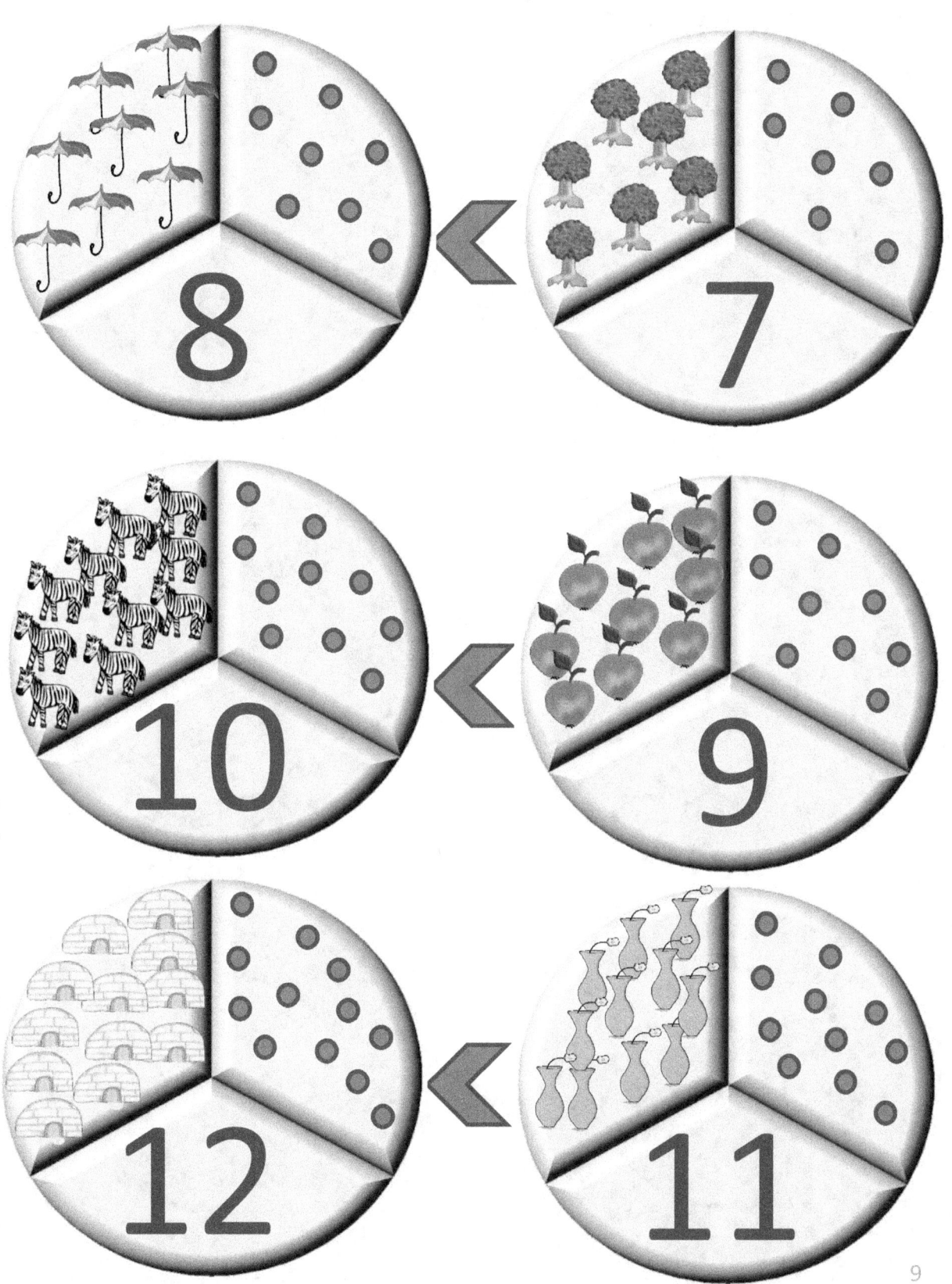

Comparison. Greater than ( > ). Fill in the missing numbers.

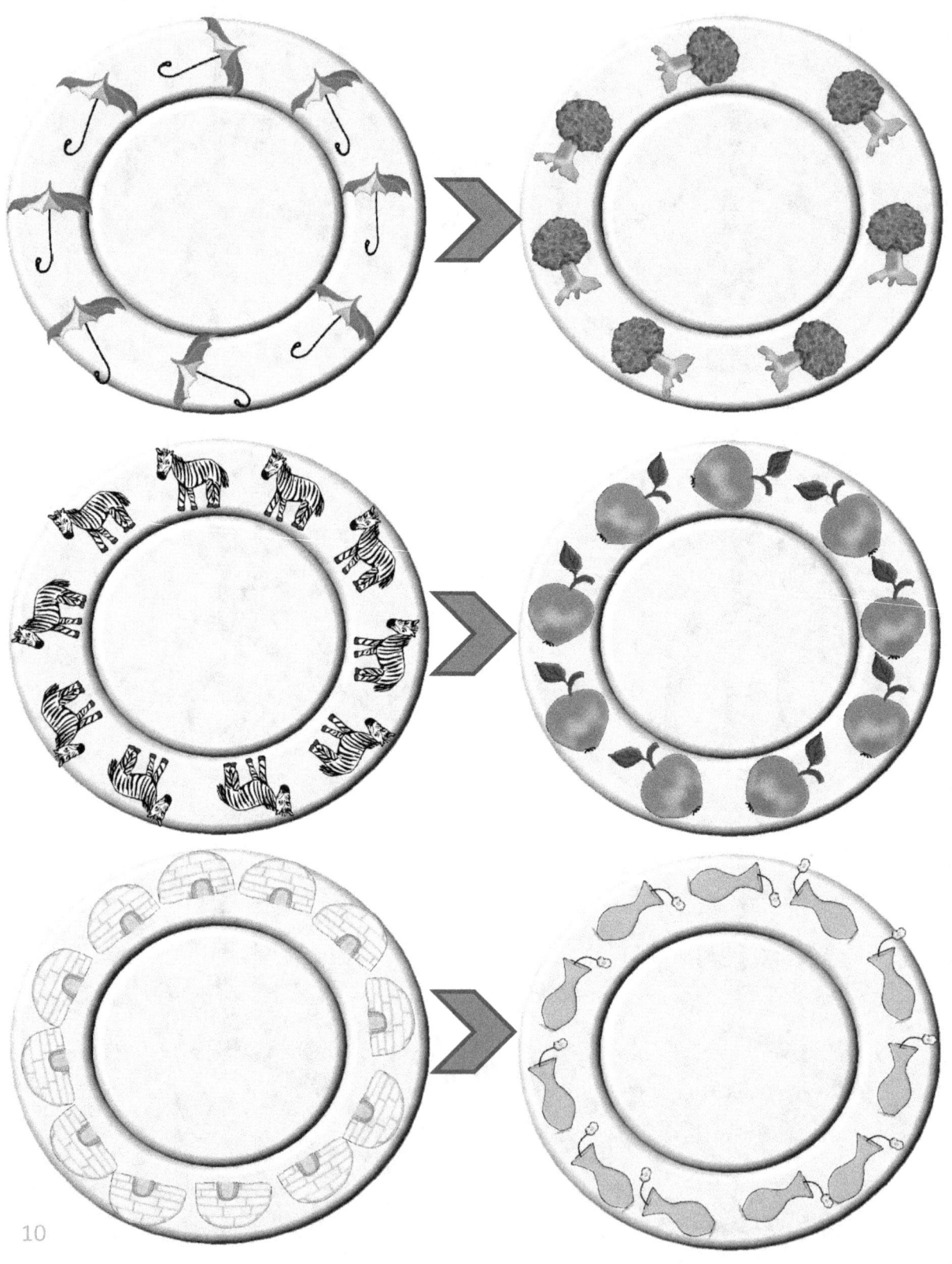

Comparison. Less than ( < ). Fill in the missing numbers.

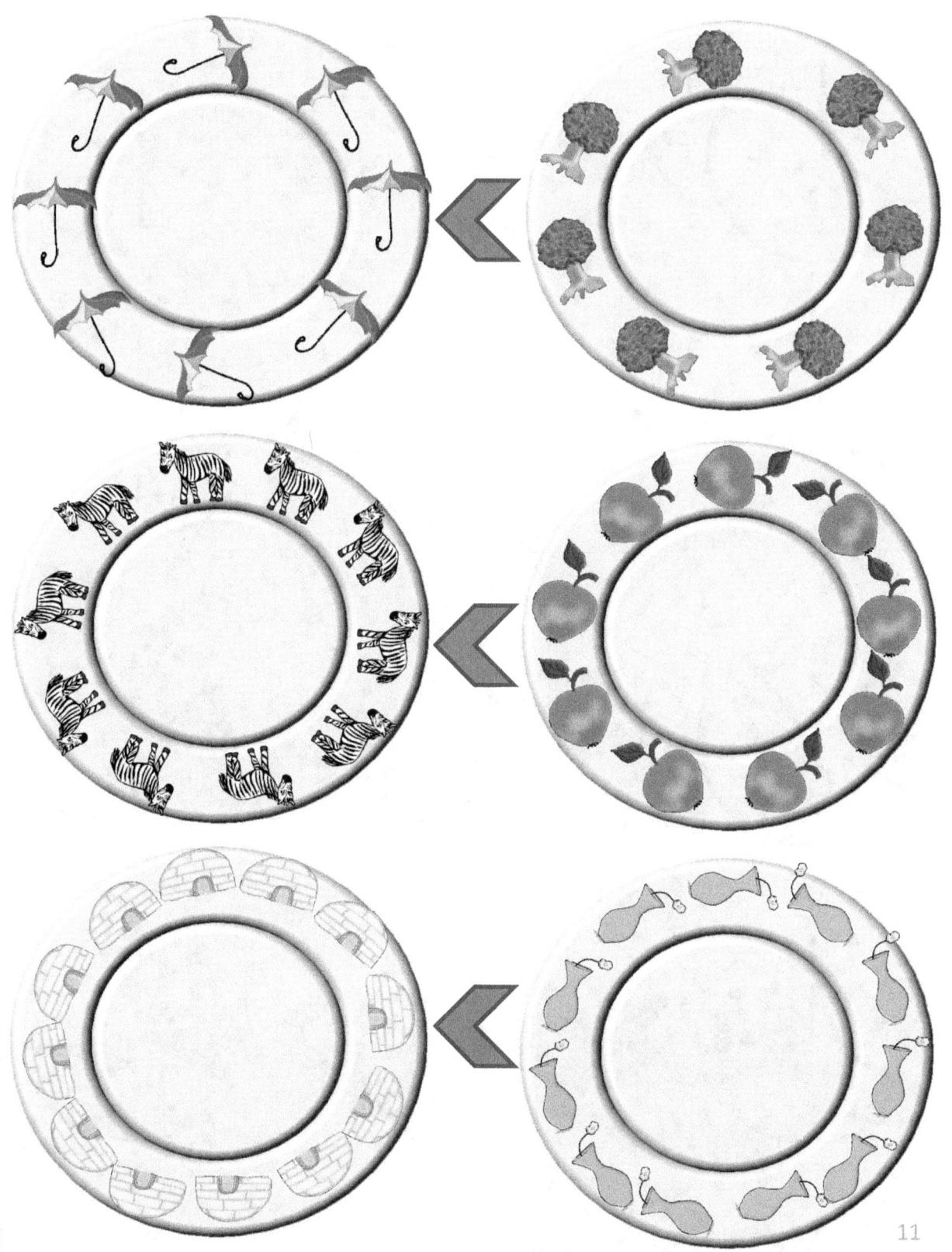

# Comparison. Greater than ( > ).

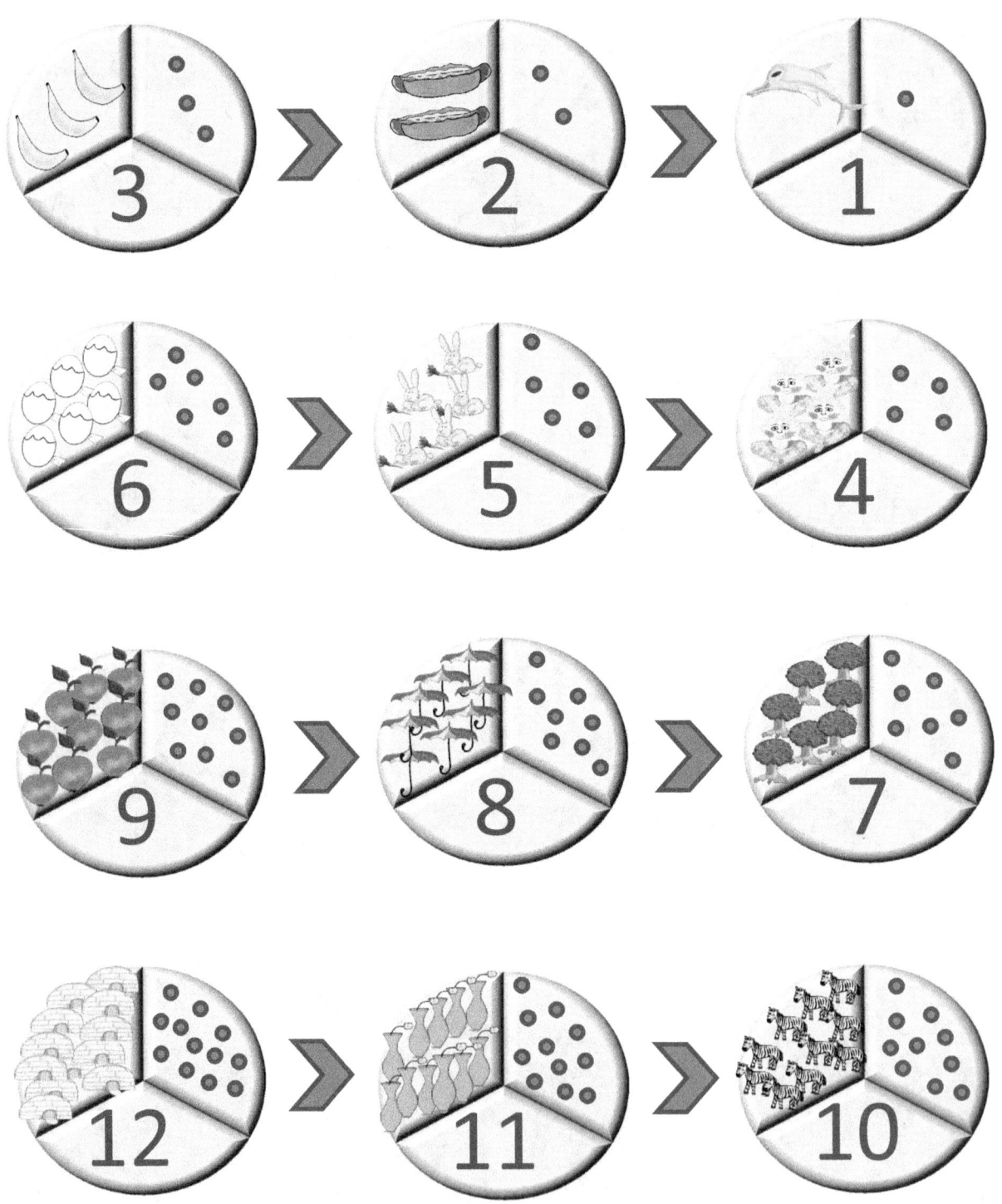

# Fill in the missing numbers.
# Greater than.

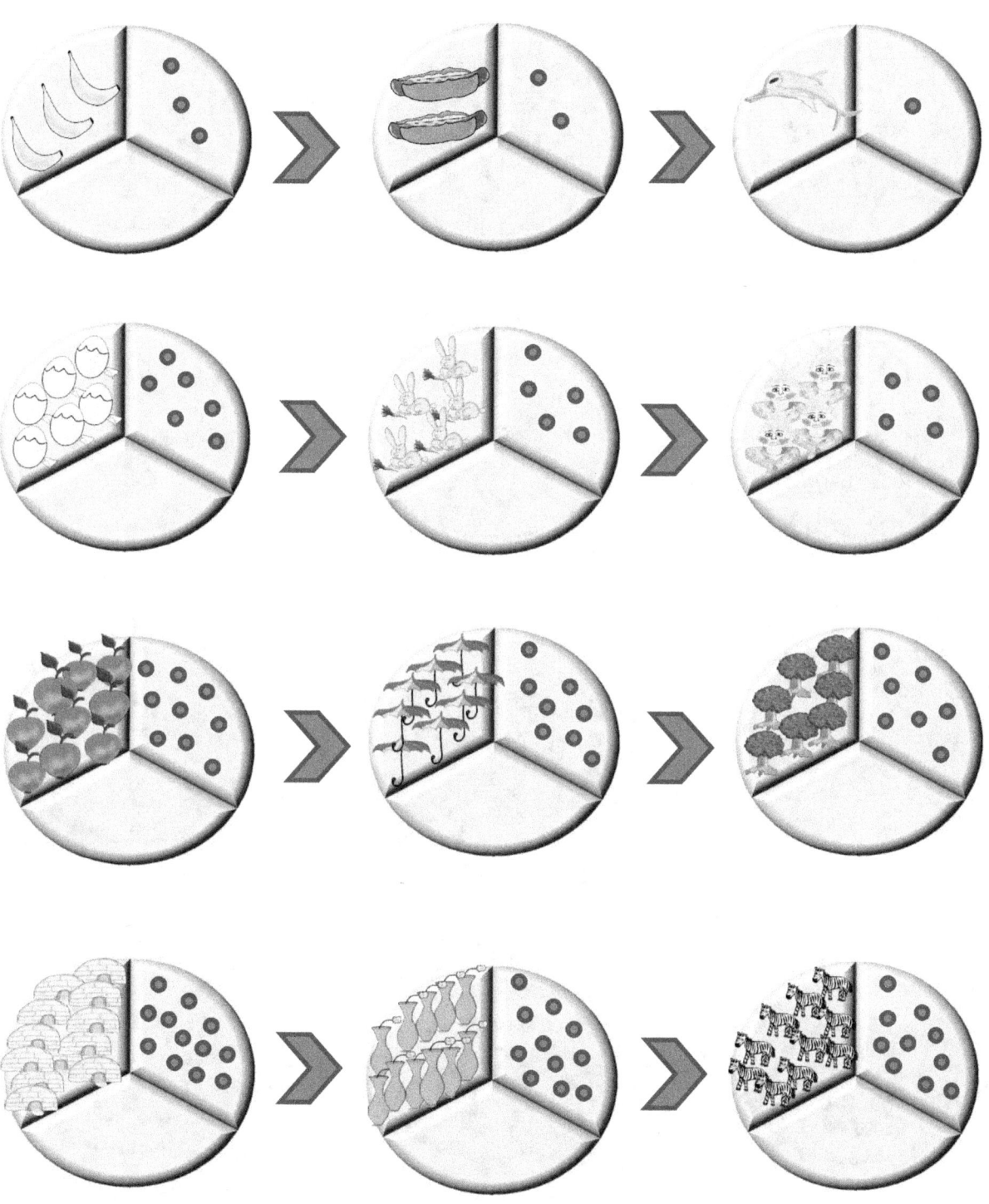

# Comparison. Less than (<).

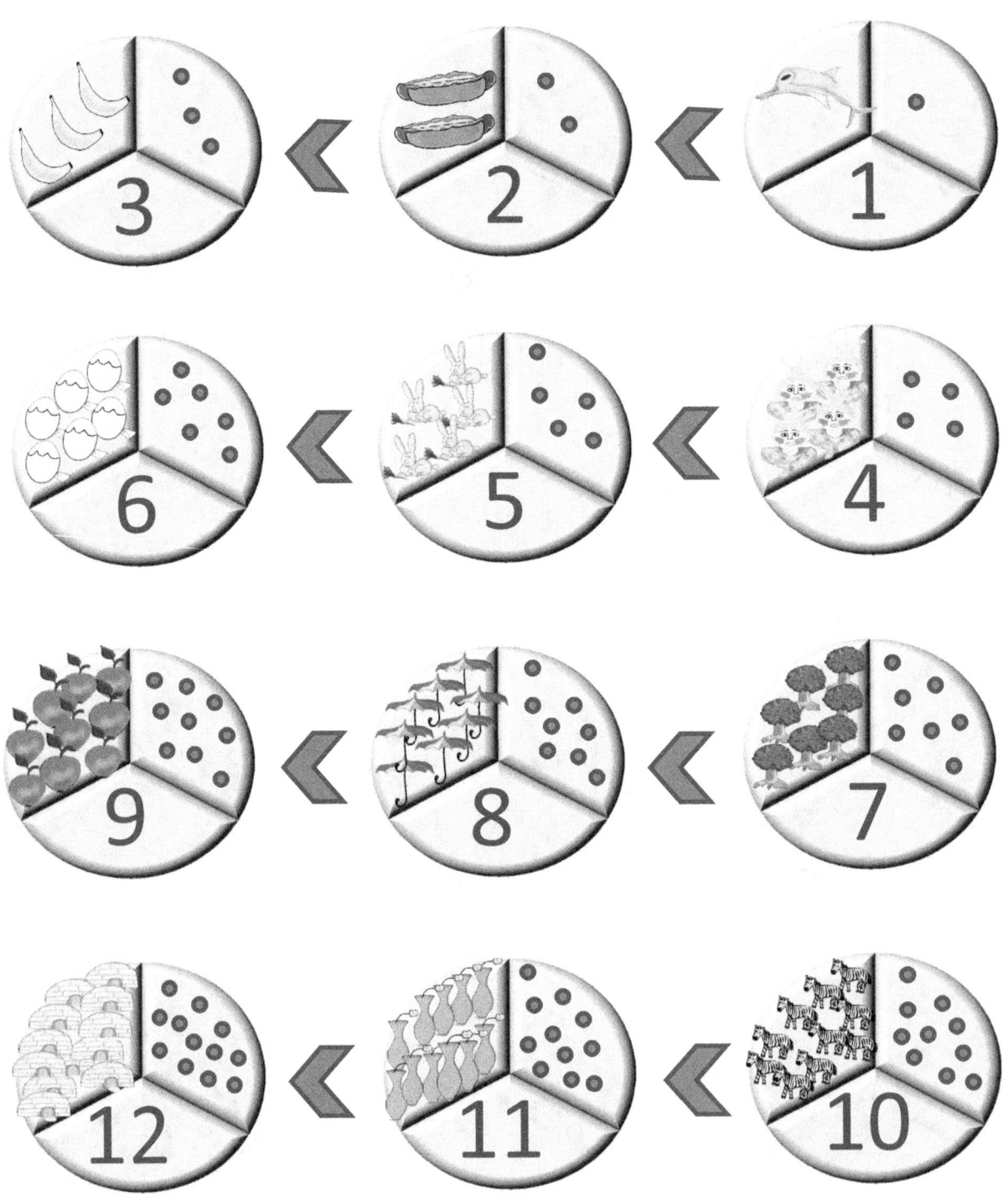

Fill in the missing numbers.
Less than.

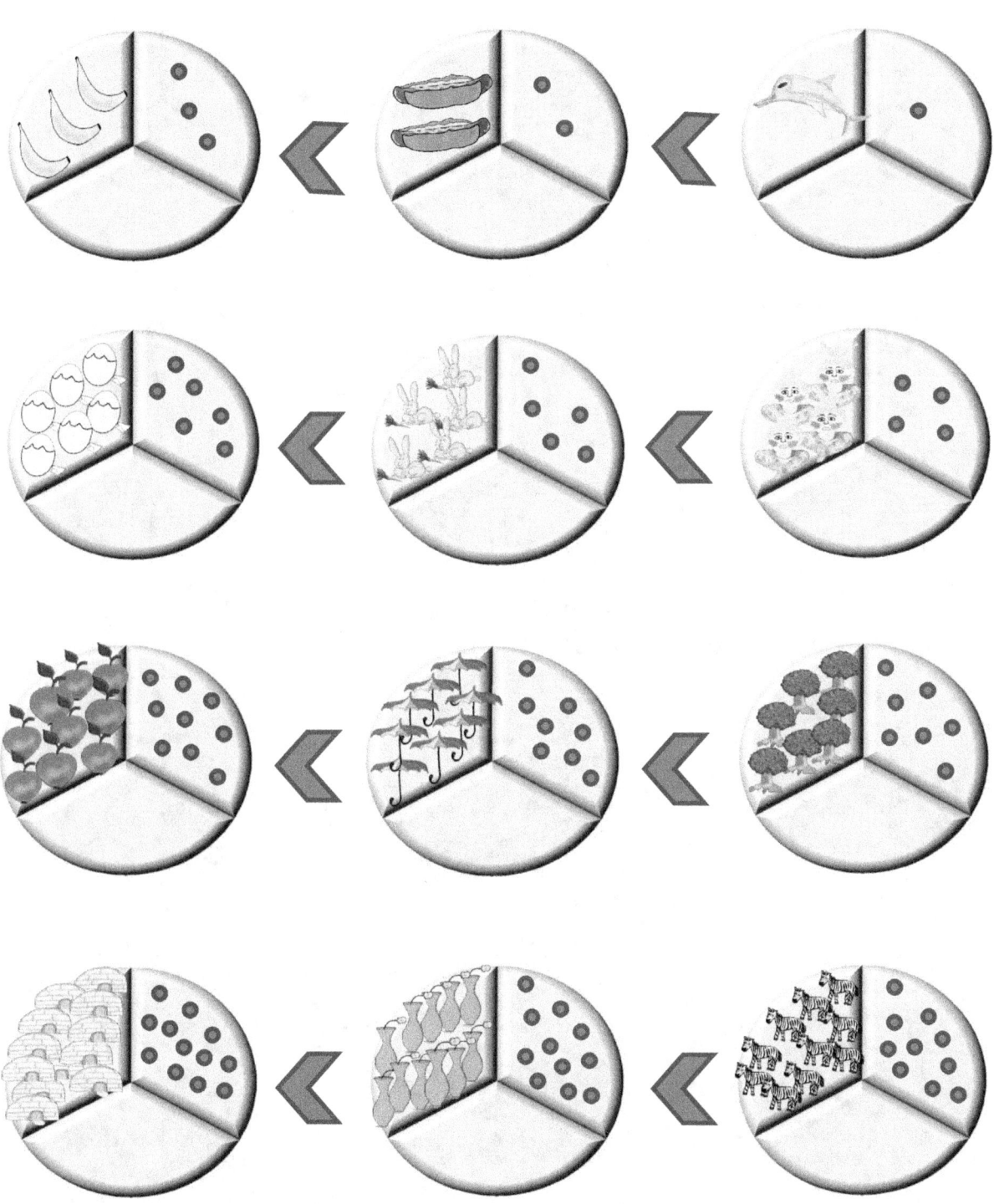

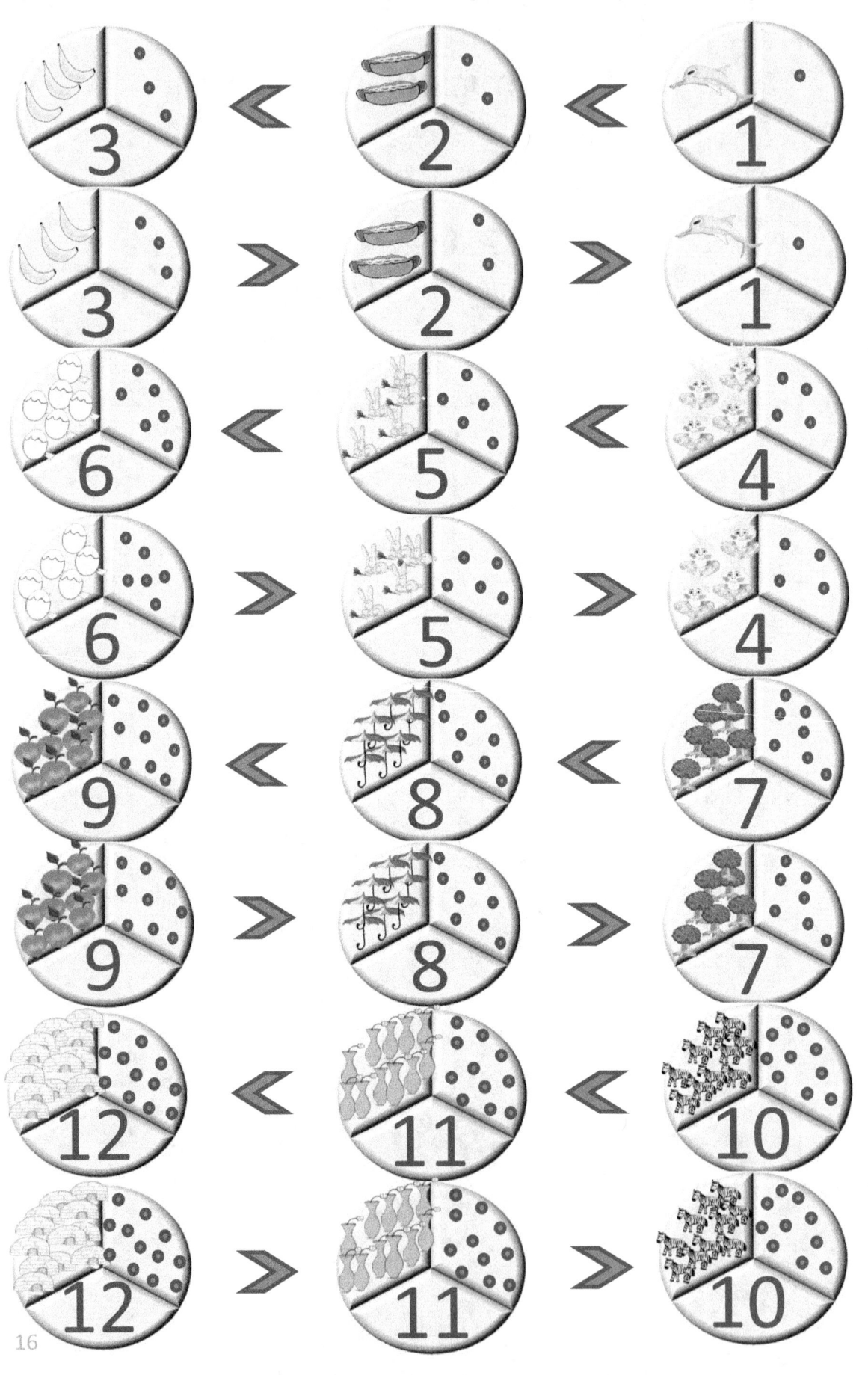

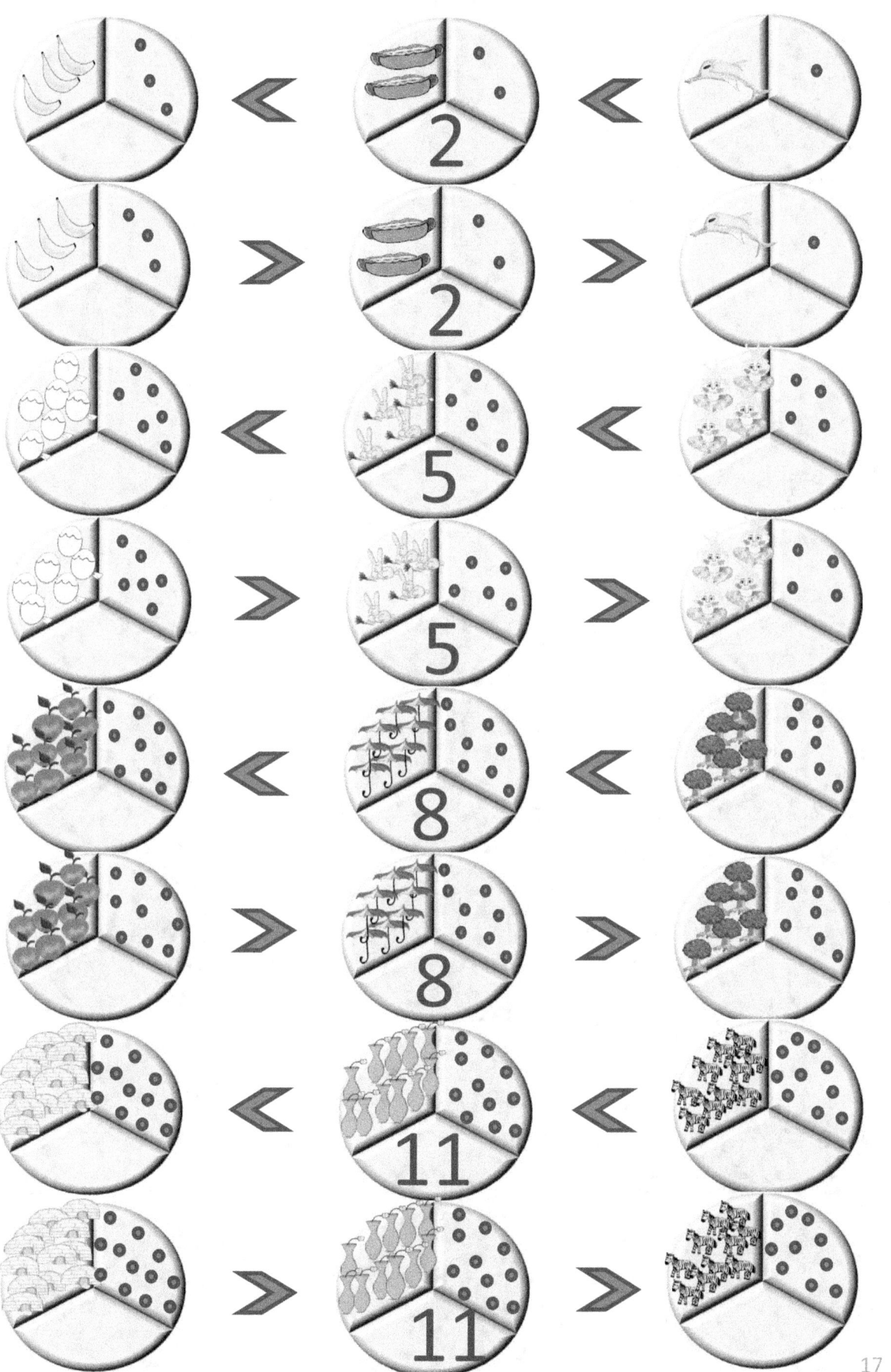

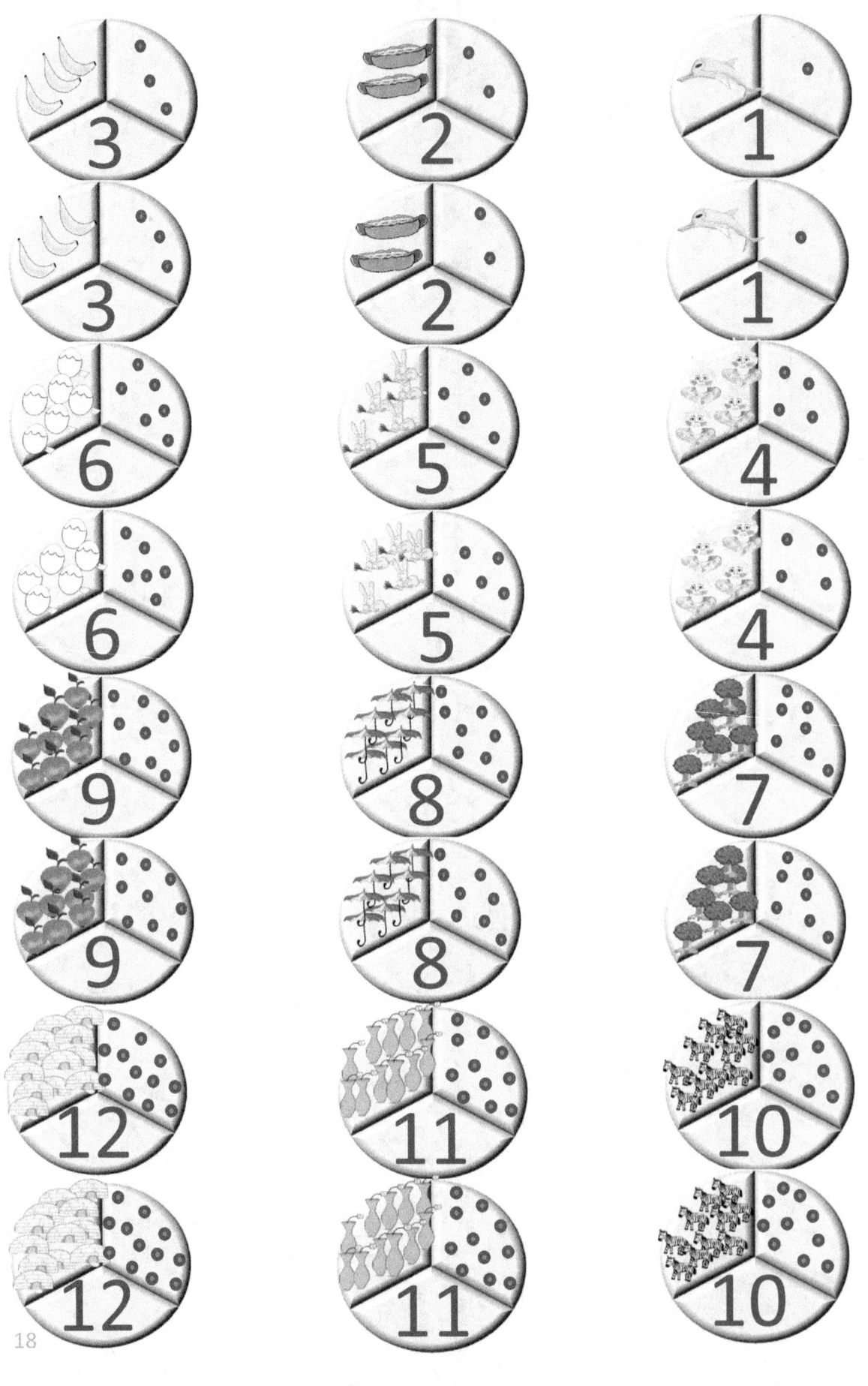

# Subtraction. Target number 1.

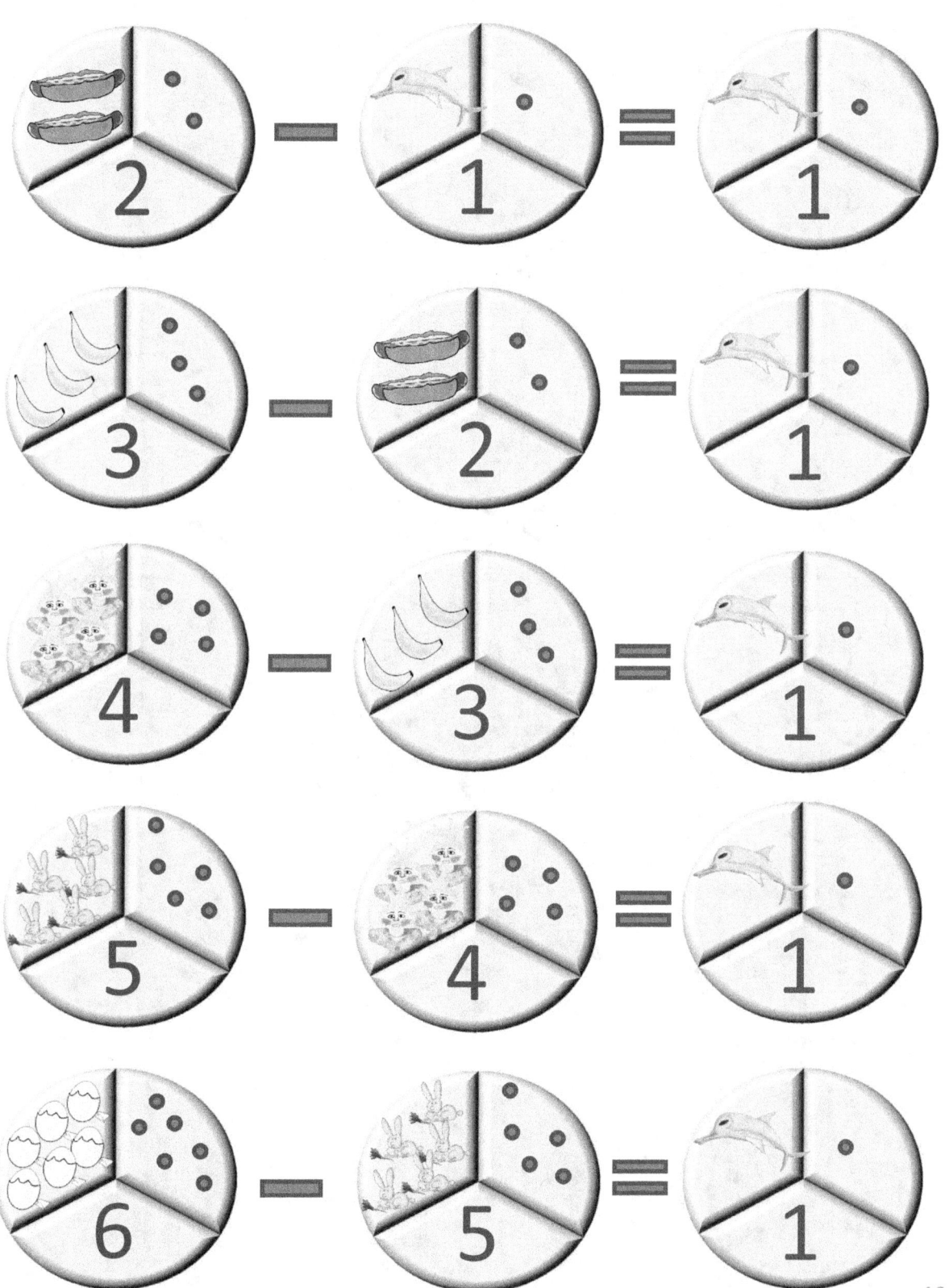

# Subtraction. Target number 1.

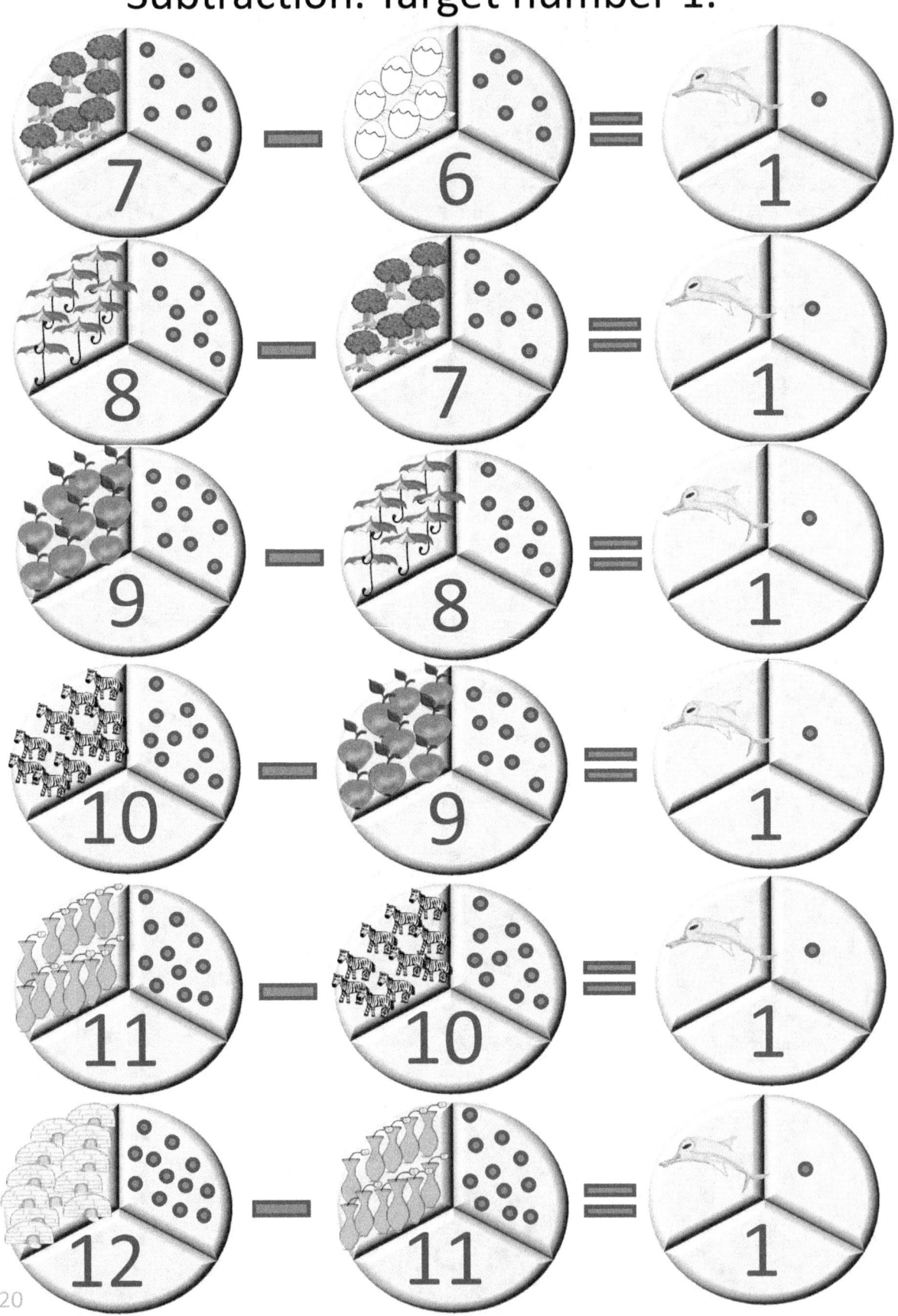

# Subtraction. Target number 1. Fill in the missing numbers.

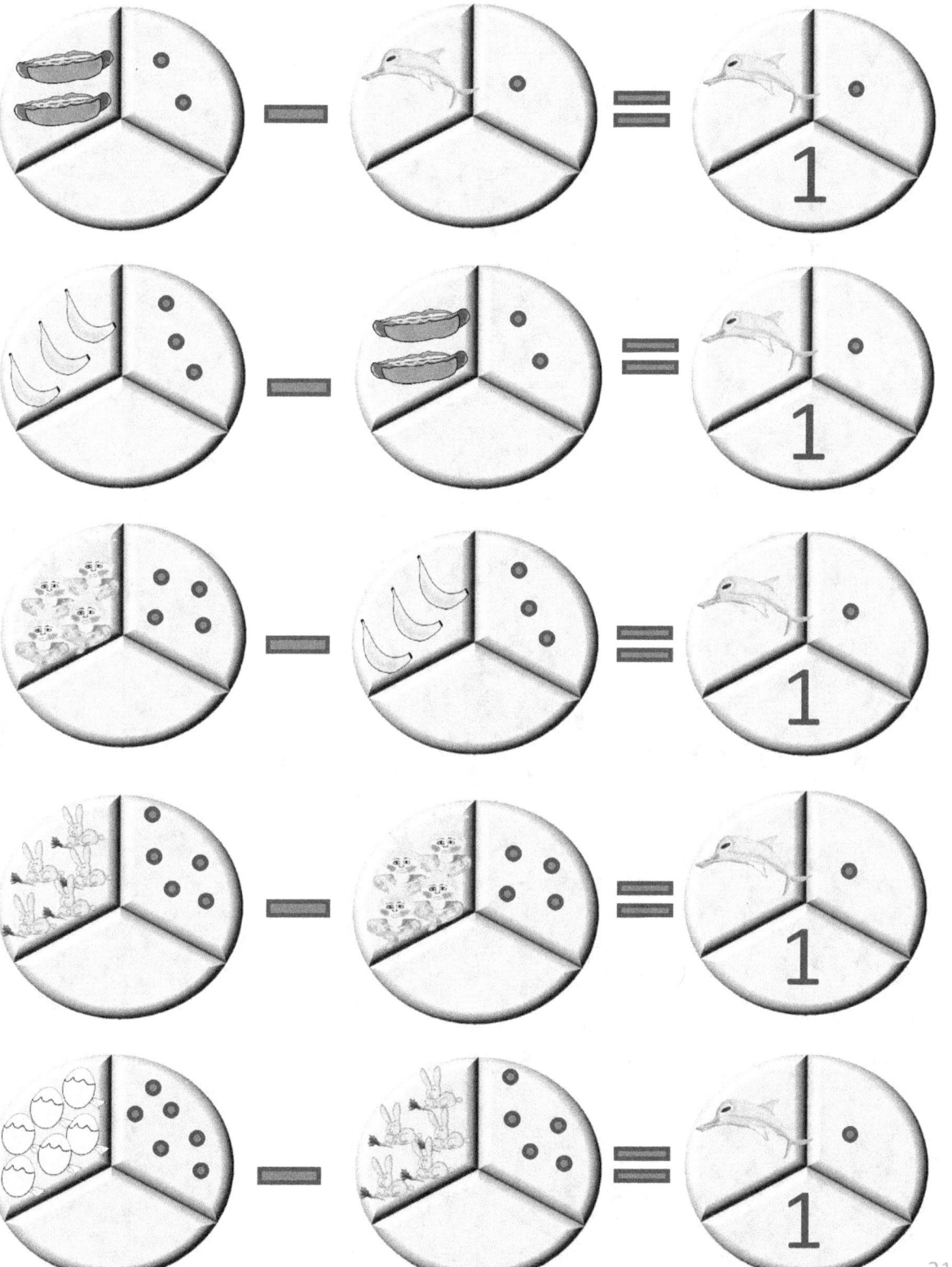

## Subtraction. Target number 1.
Fill in the missing numbers.

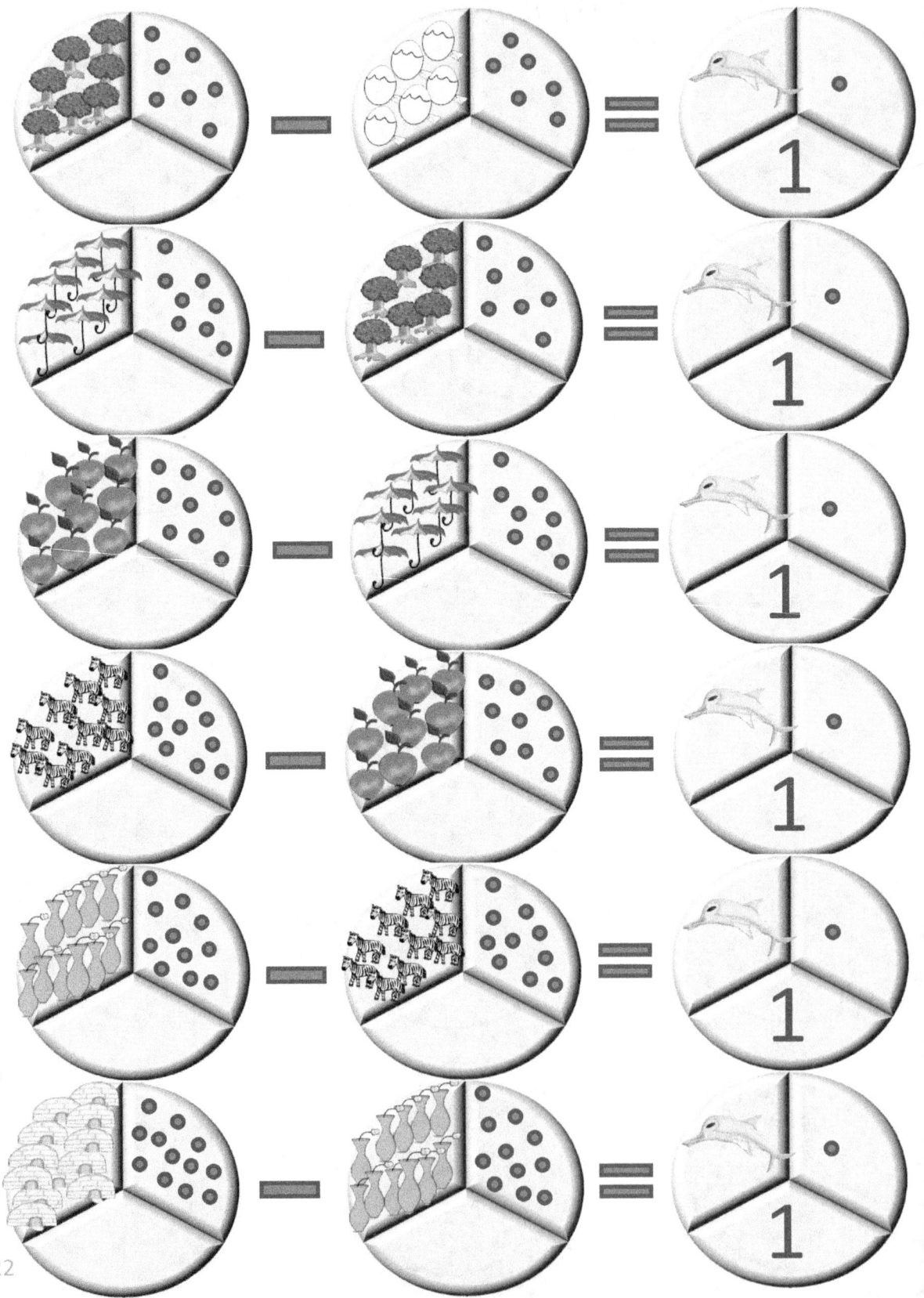

# Subtraction. Target number 2.

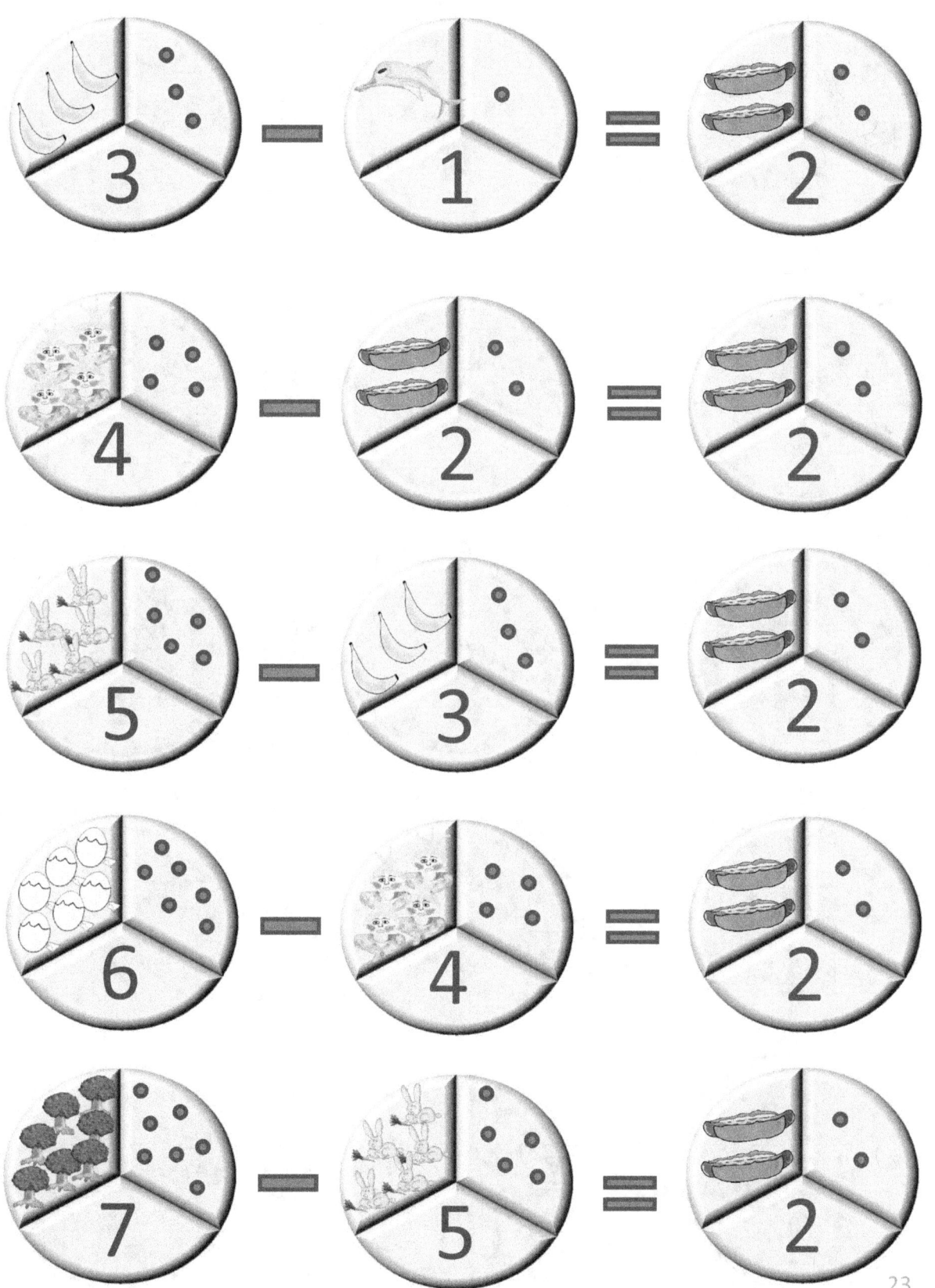

# Subtraction. Target number 2.

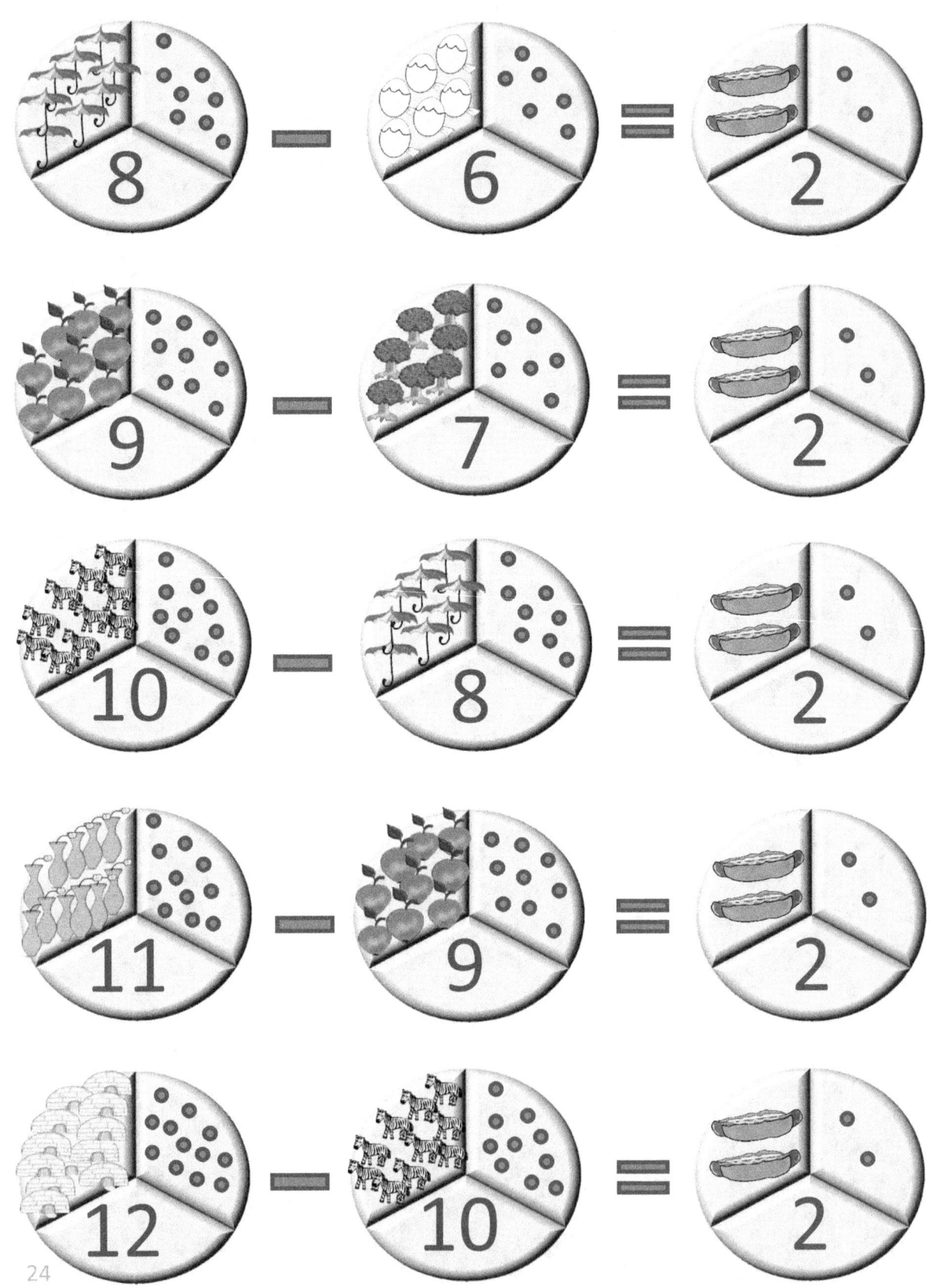

# Subtraction. Target number 2. Fill in the missing numbers.

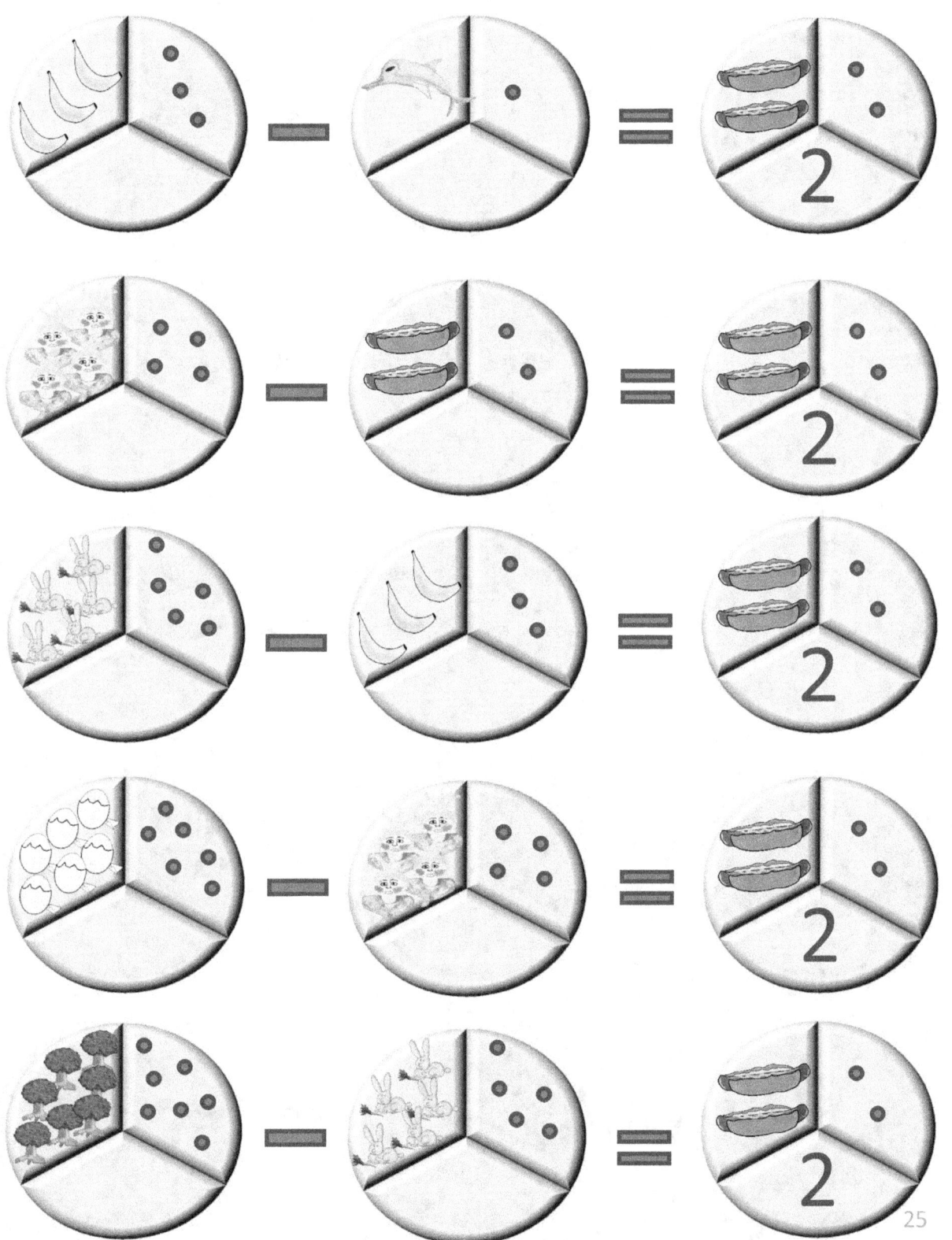

# Subtraction. Target number 2. Fill in the missing numbers.

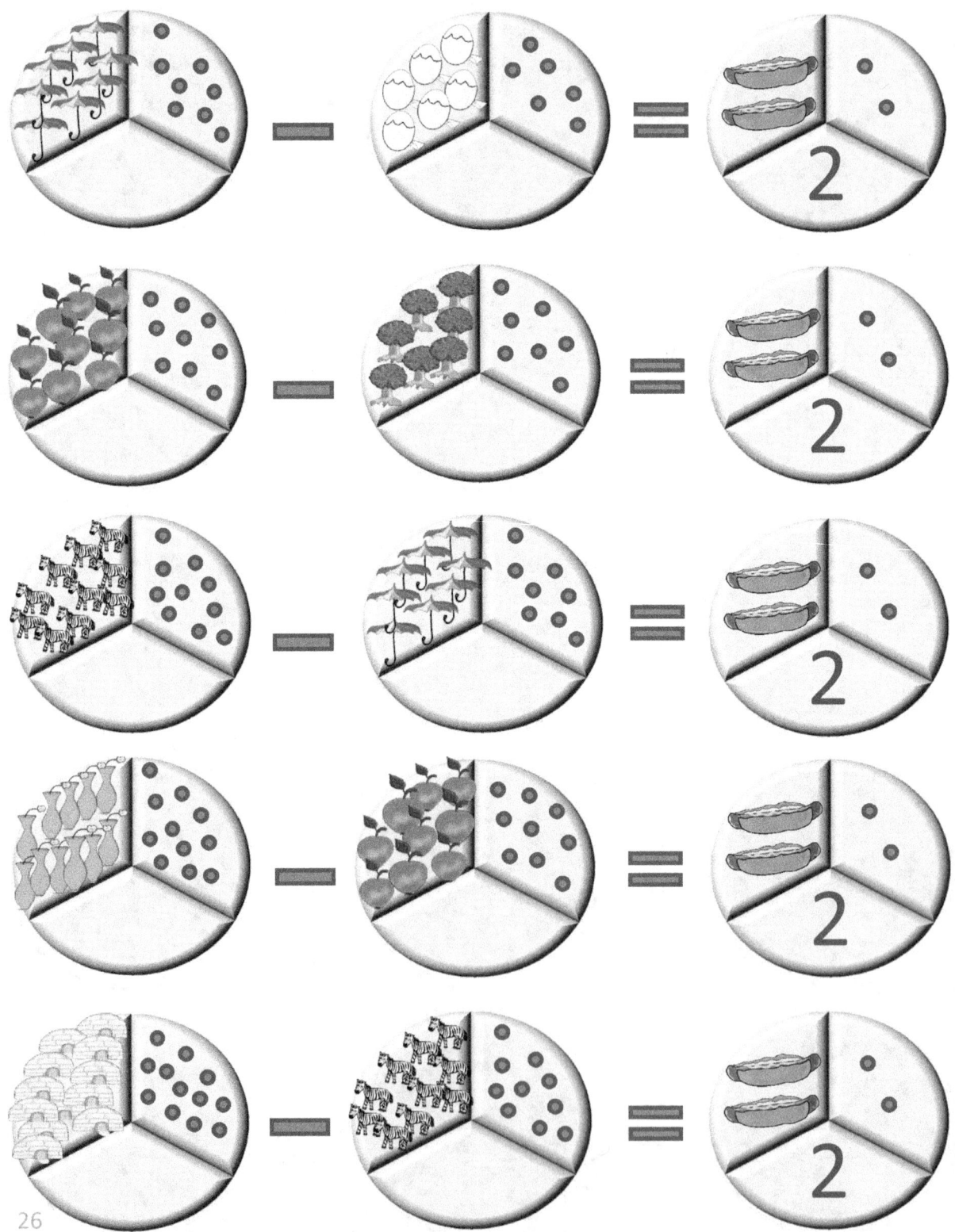

# Multiplication x 1

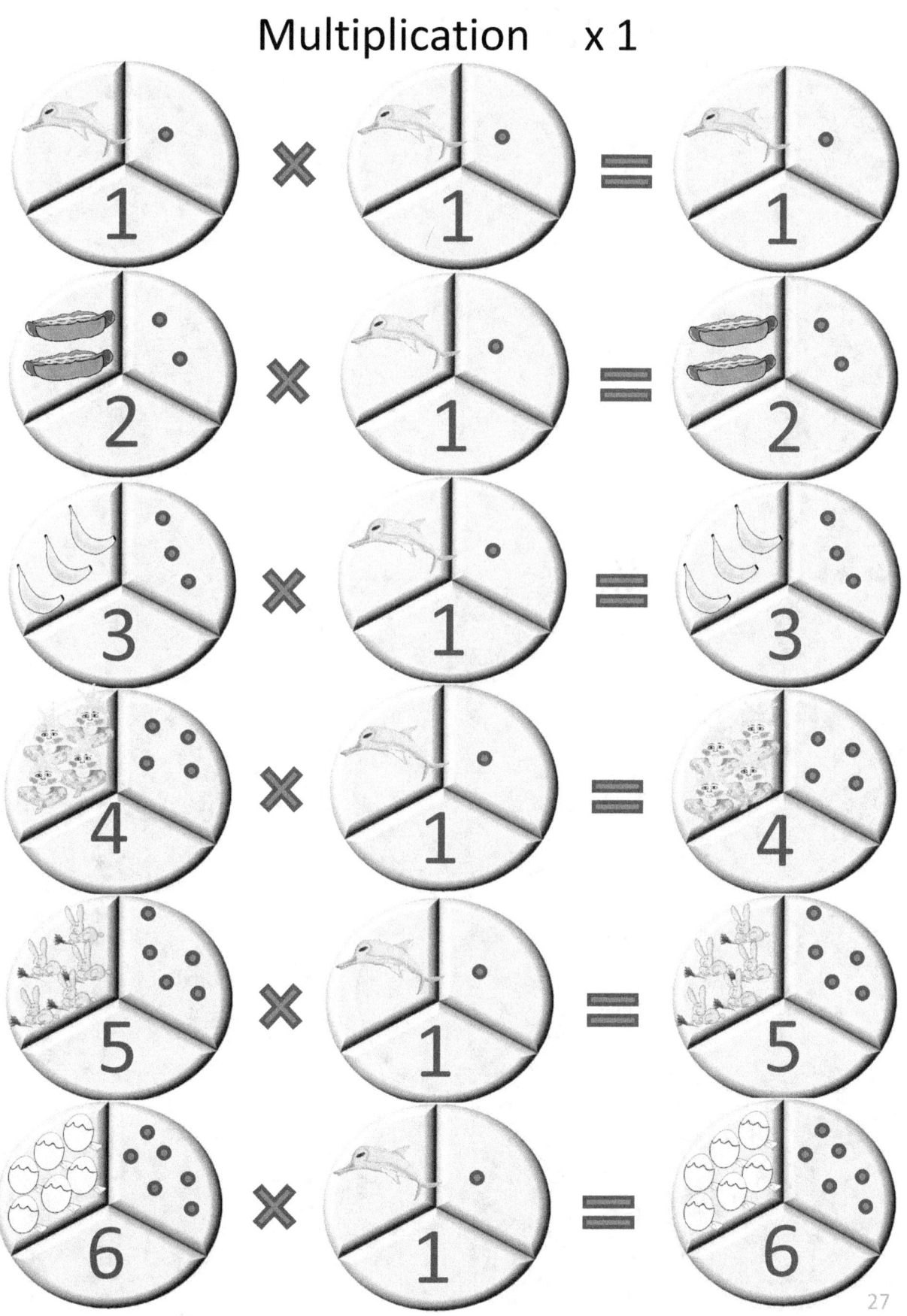

# Multiplication x 1

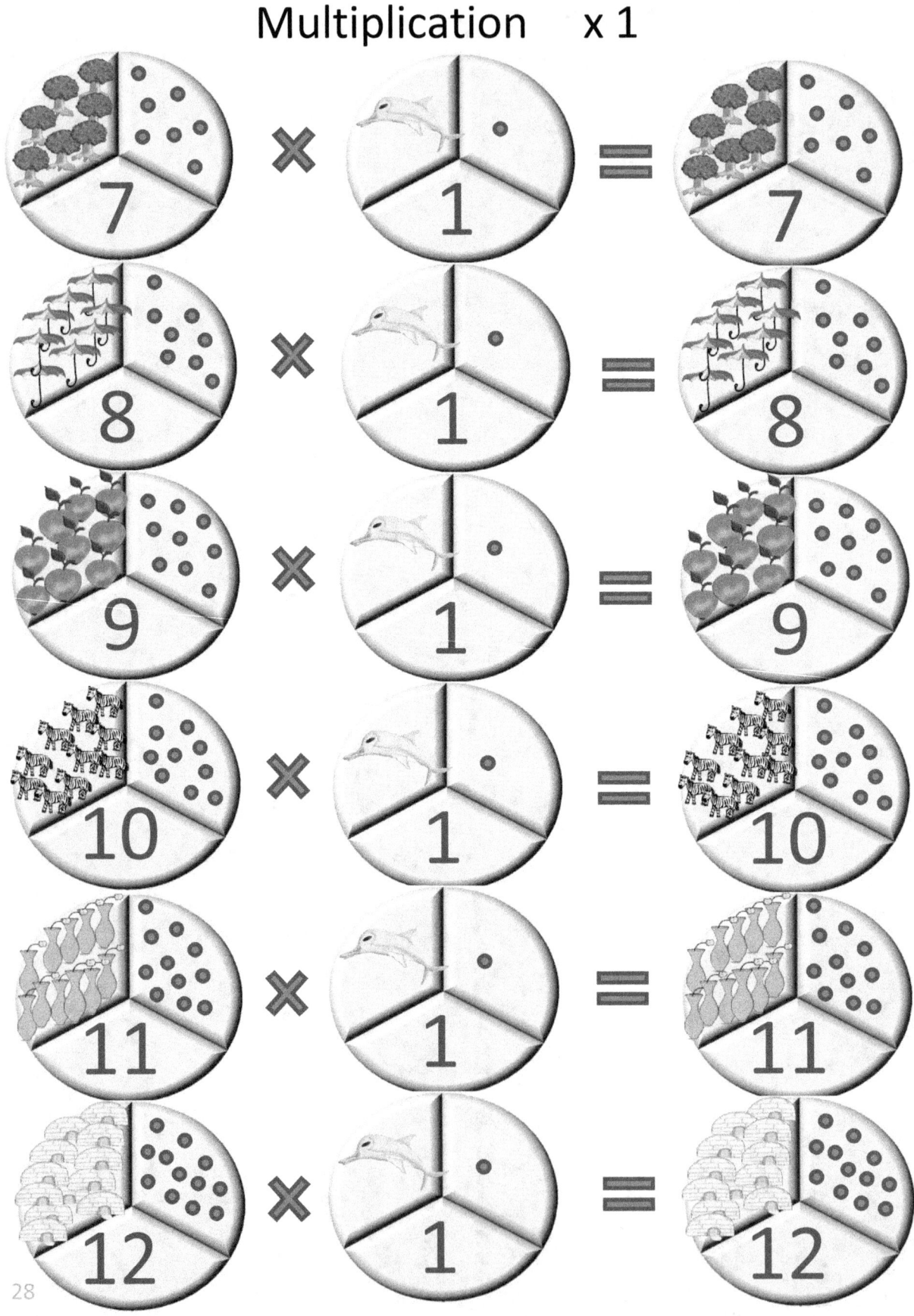

# Multiplication x 1. Fill in the missing numbers.

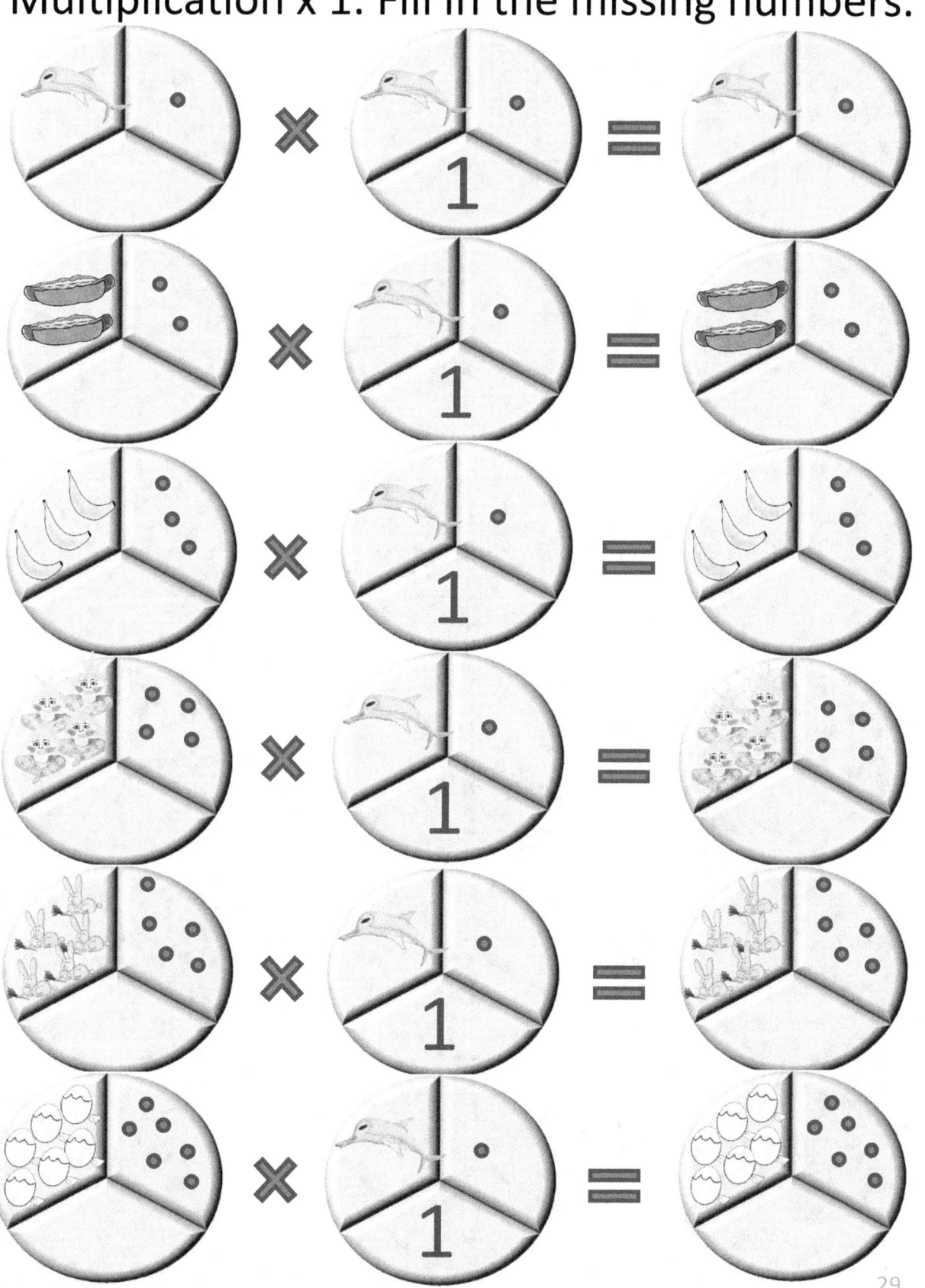

# Multiplication x 1. Fill in the missing numbers.

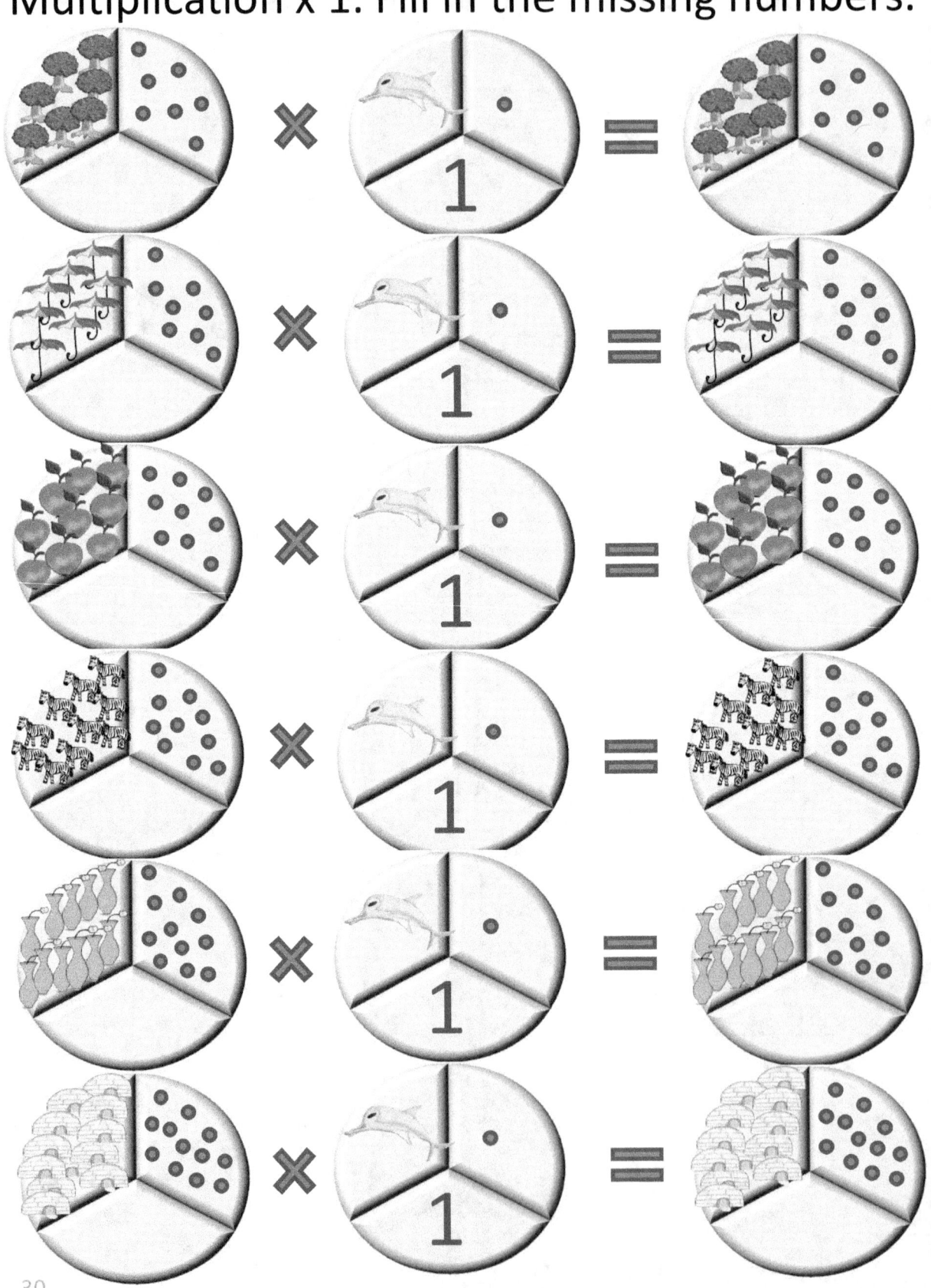

# Multiplication x 2

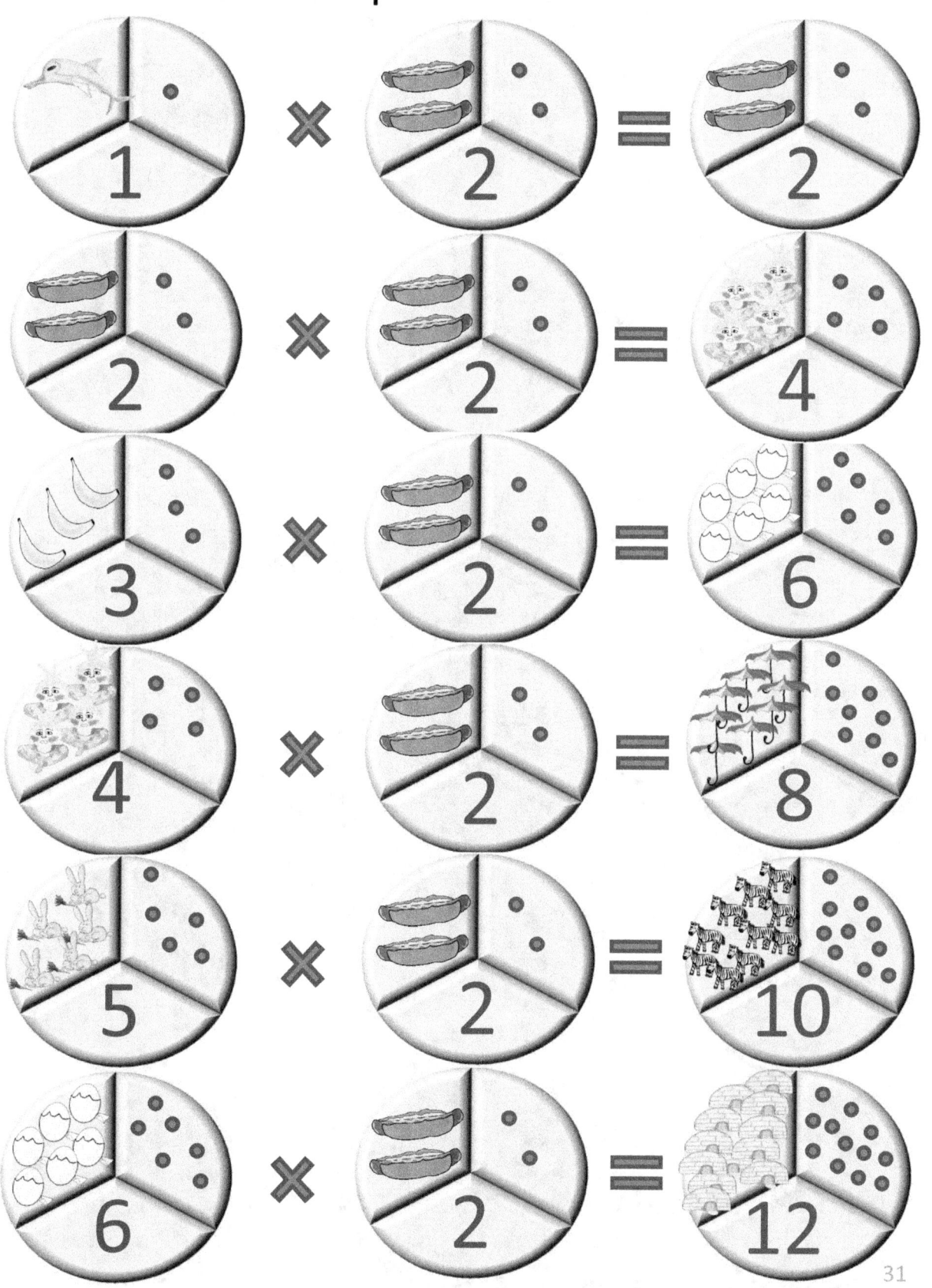

# Multiplication x 2. Fill in the missing numbers.

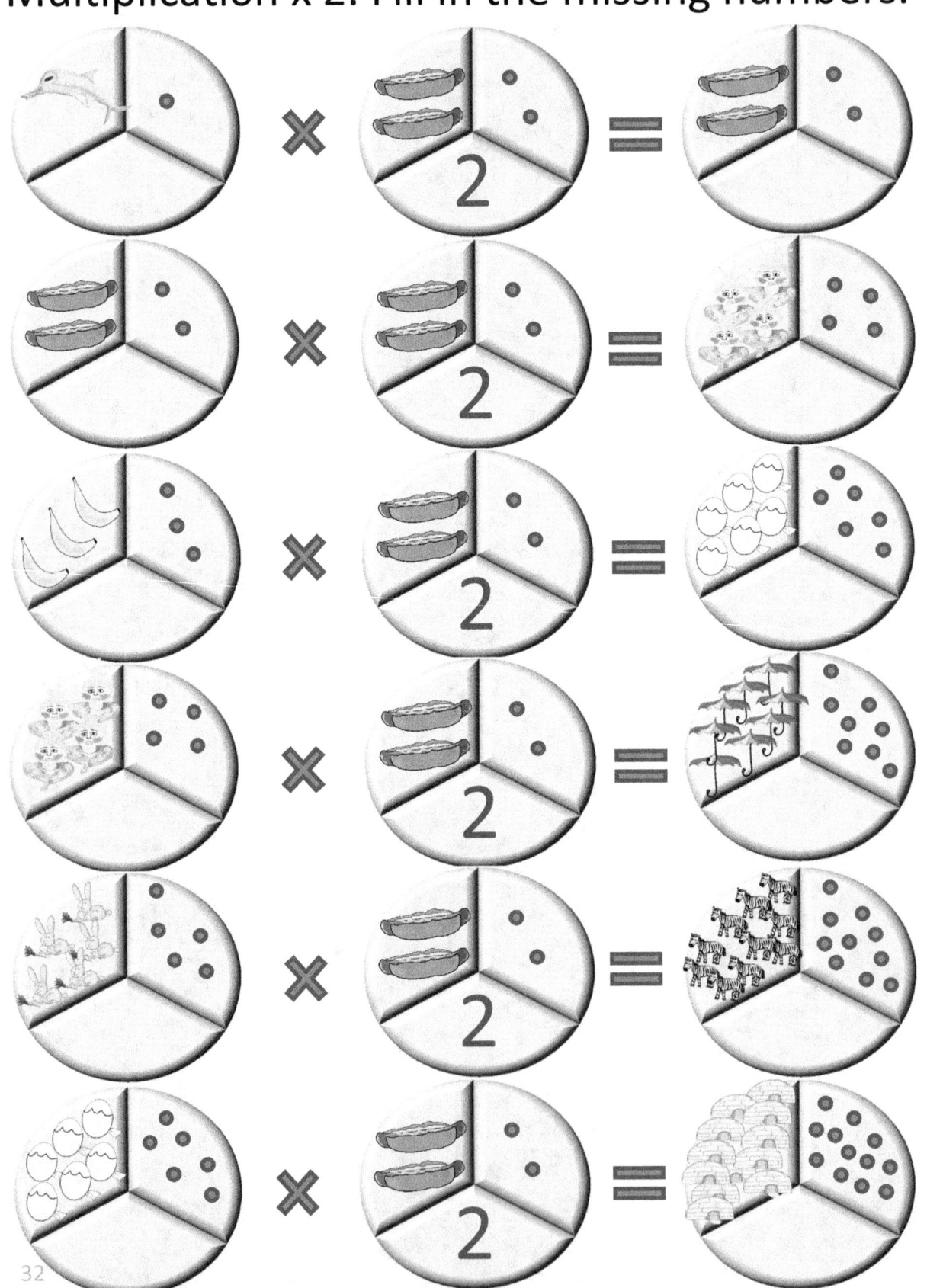

# Multiplication  x 3

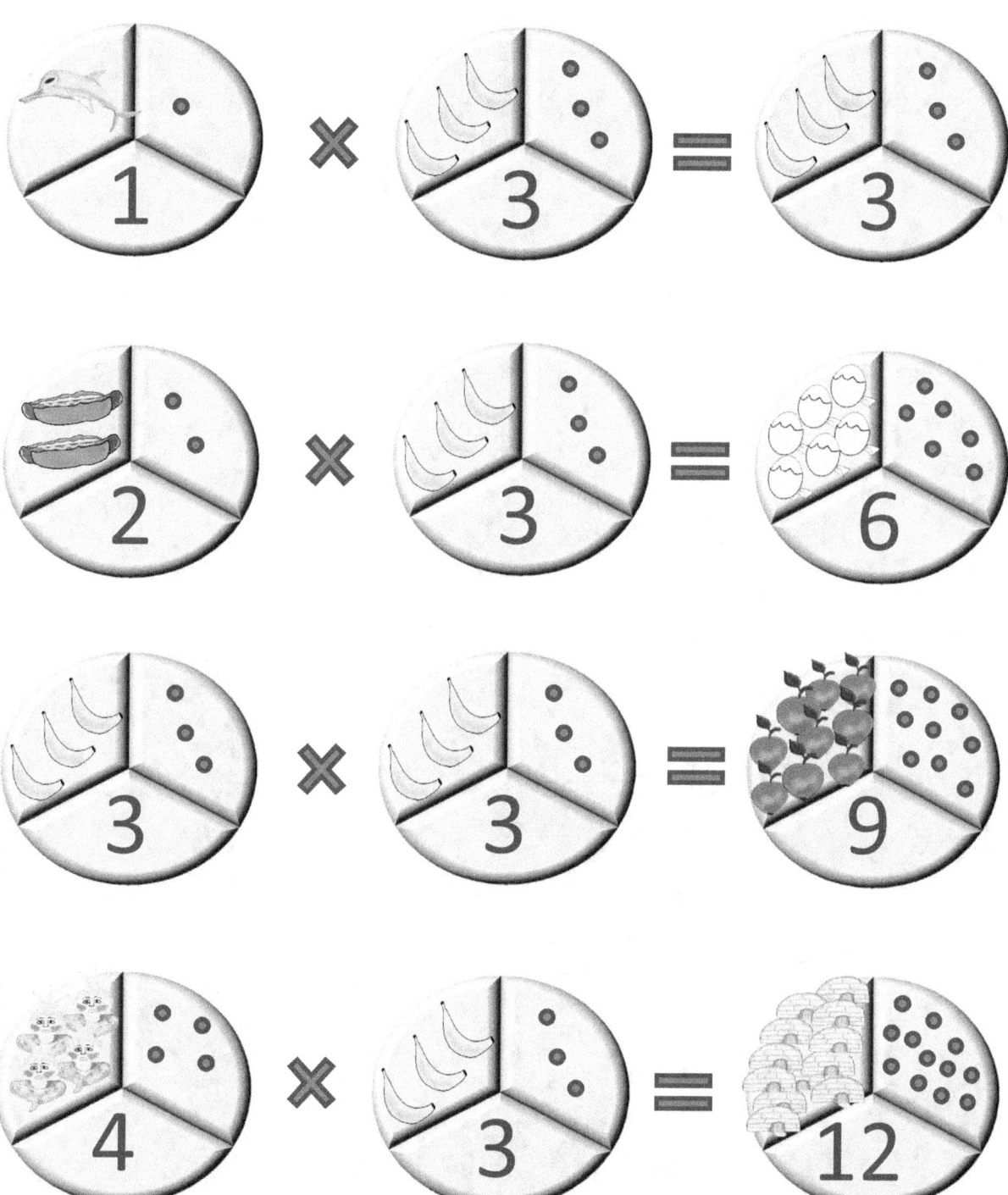

# Multiplication x 3. Fill in the missing numbers.

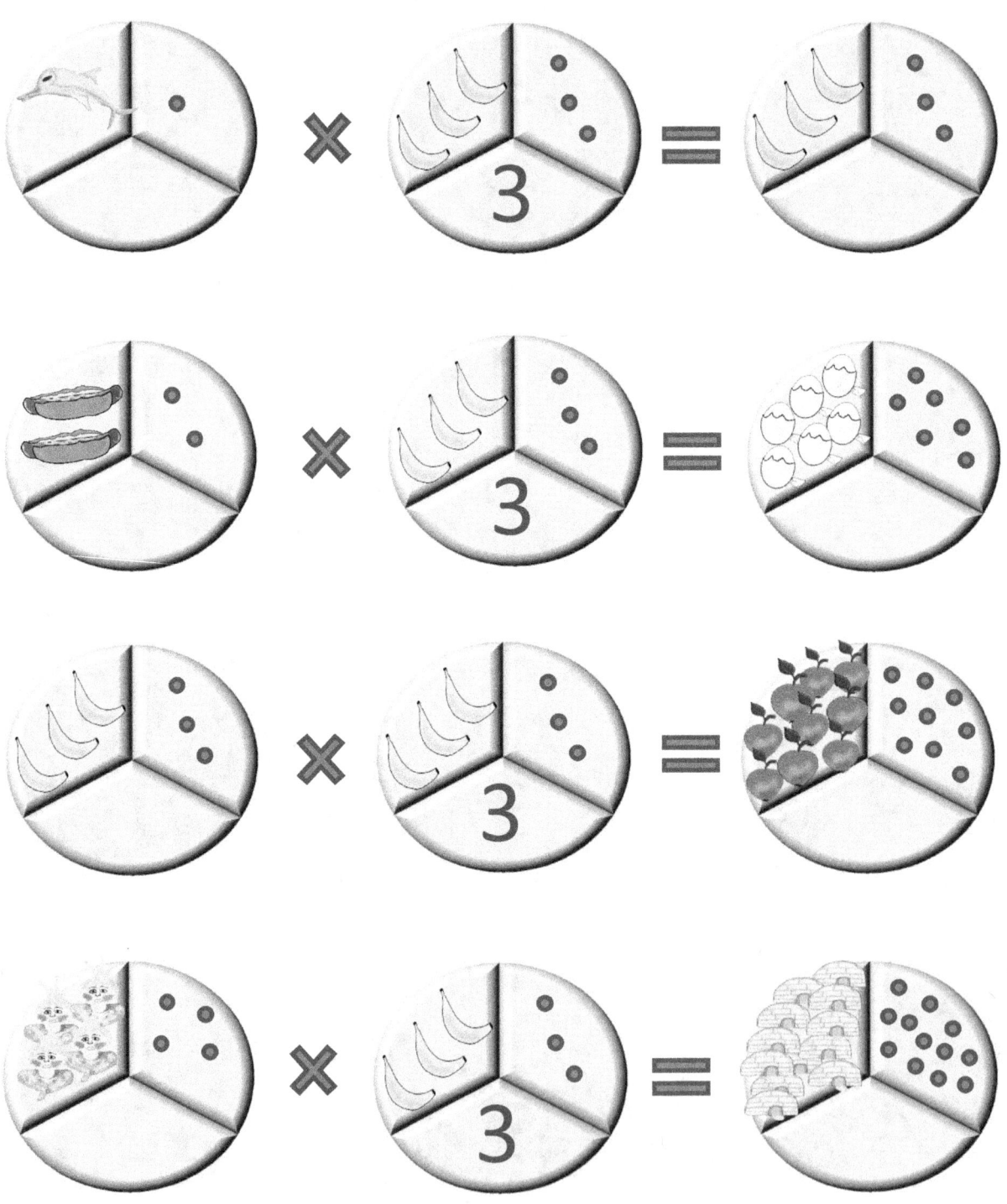

# Multiplication x 4

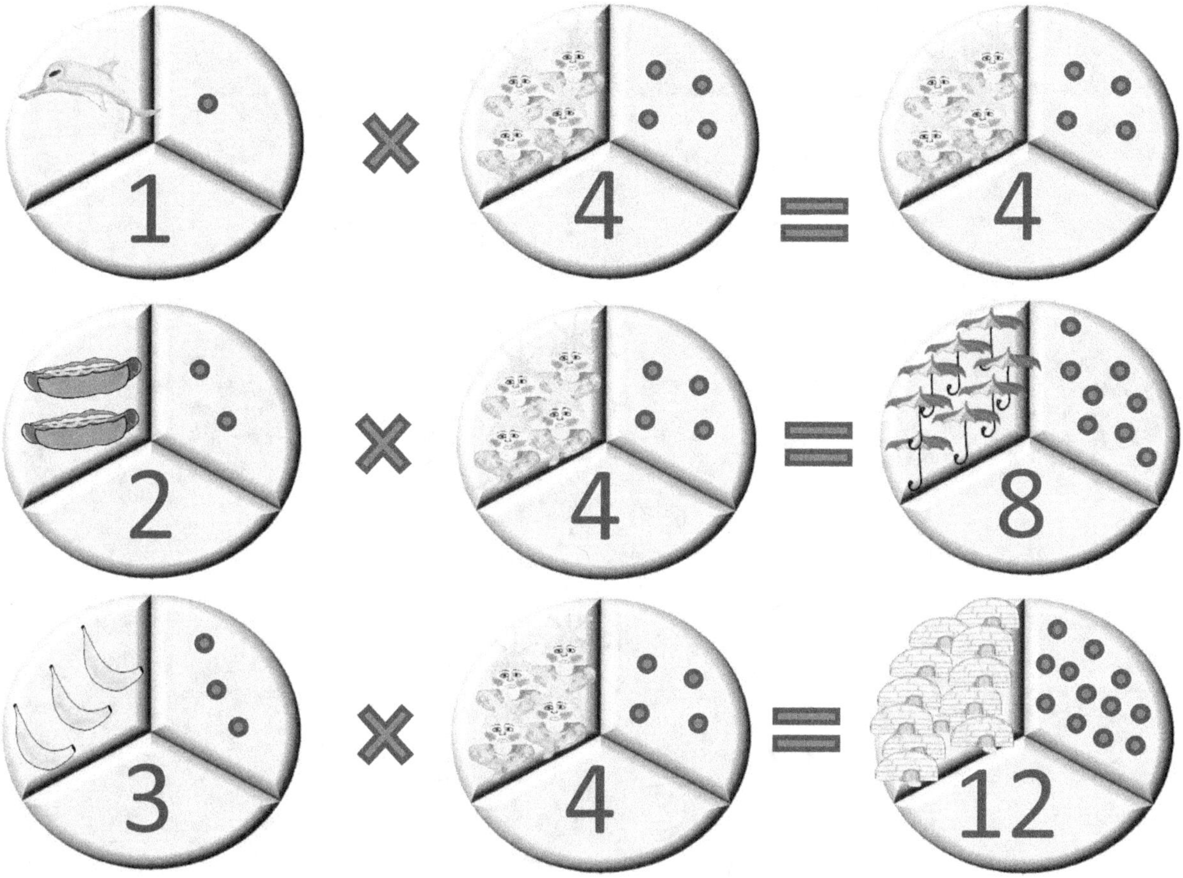

# Multiplication x 5

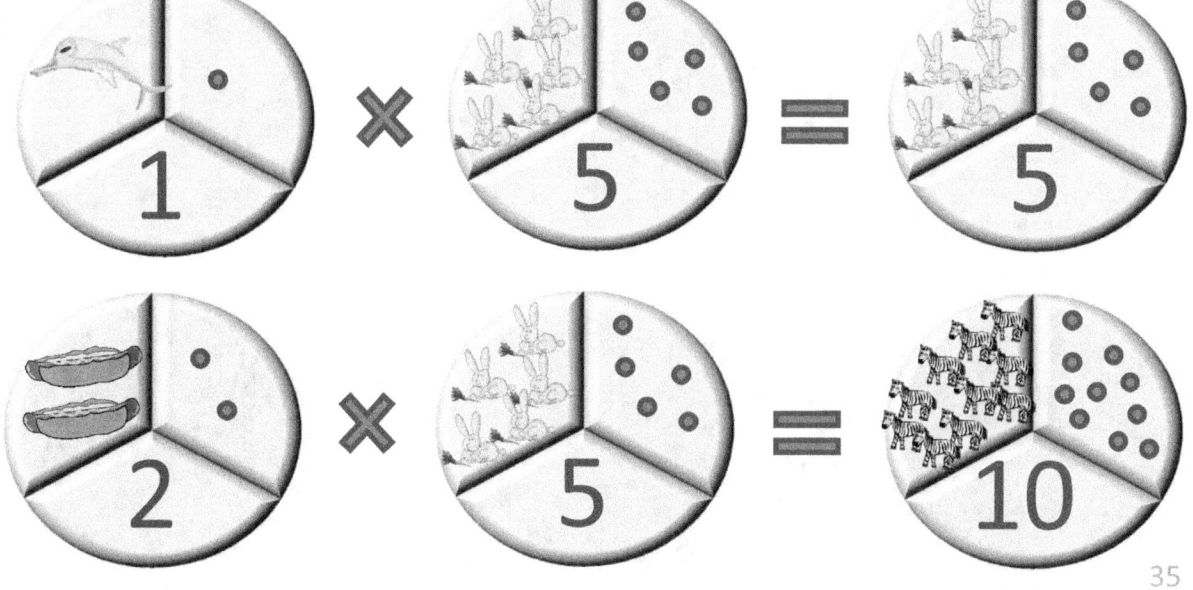

# Multiplication x 4. Fill in the missing numbers.

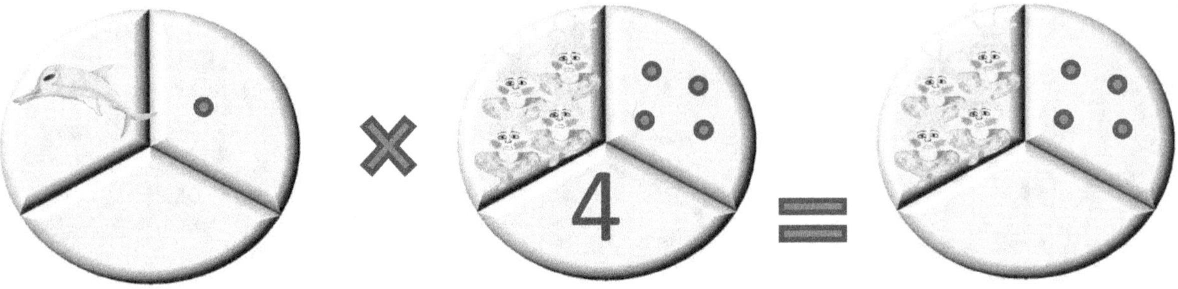

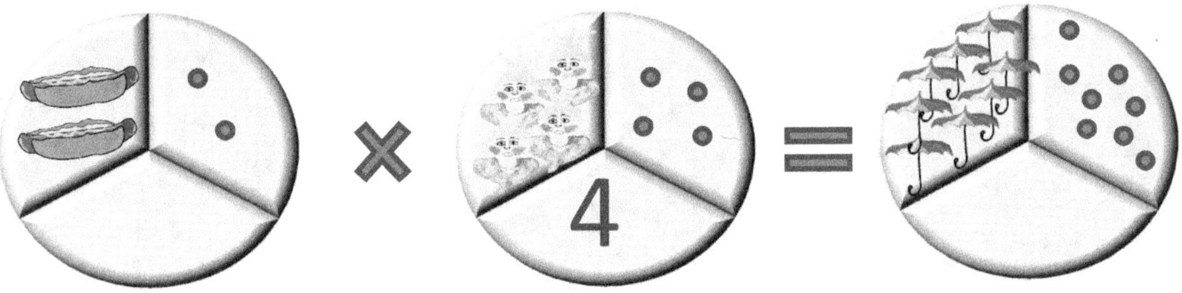

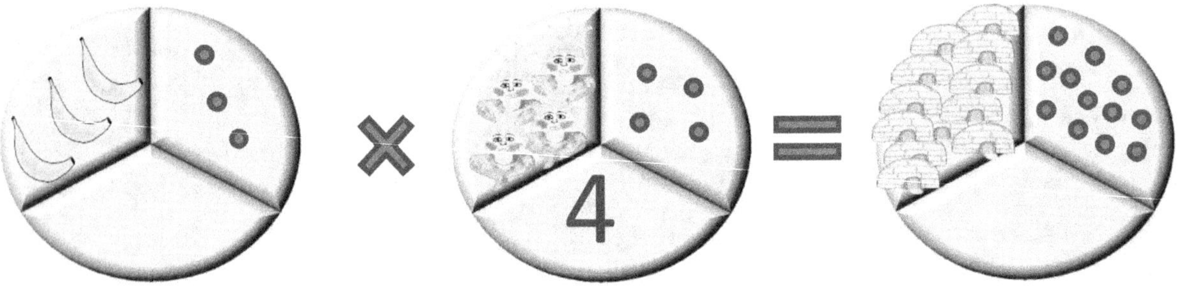

# Multiplication x 5. Fill in the missing numbers.

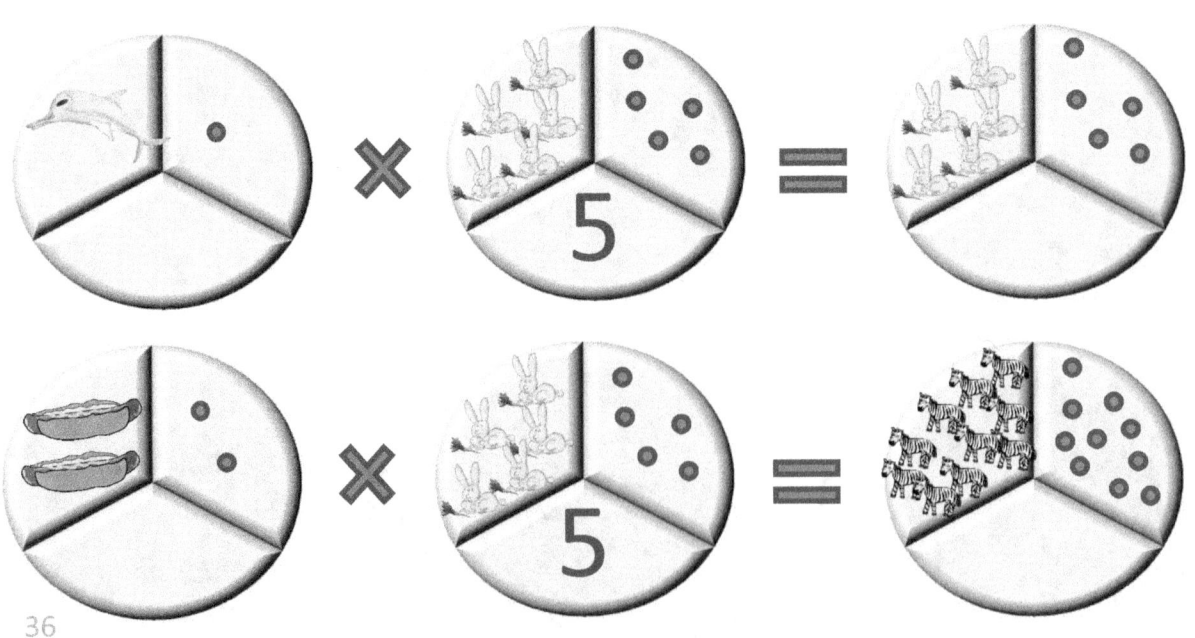

www.ingramcontent.com/pod-product-compliance
Lightning Source LLC
Chambersburg PA
CBHW062345220526
45469CB00008B/2838